Religion

For centuries, scripture and theology were the focus of prodigious amounts of scholarship and publishing, dominated in the English-speaking world by the work of Protestant Christians. Enlightenment philosophy and science, anthropology, ethnology and the colonial experience all brought new perspectives, lively debates and heated controversies to the study of religion and its role in the world, many of which continue to this day. This series explores the editing and interpretation of religious texts, the history of religious ideas and institutions, and not least the encounter between religion and science.

A Series of Discourses on the Christian Revelation

In 1817 he Scottish mathematician and churchman Thomas Chalmers (1780-1847), who was later invited to write one of the Bridgewater Treatises (also reissued in this series) published this book, based on weekday sermons preached by him in Glasgow. His main aim is to refute, or at least 'soften and subdue' the 'infidel' argument that because the earth and humanity are such insignificant parts of the universe, God - if he existed - would not care about them. However, he is also addressing the 'narrow and intolerant professors' who 'take an alarm' at the idea of philosophy rather than incorporating science into their Christian preaching. Chalmers writes from the viewpoint of an admirer of science and modern astronomy, agrees that 'space knows no termination' and admits to the possibility of extraterrestrial life on 'greater globes than ours'. However, he also argues that wonder at the magnificence of creation and even at God's work is not enough, and that a truly moral Christian life is essential for salvation.

Cambridge University Press has long been a pioneer in the reissuing of out-of-print titles from its own backlist, producing digital reprints of books that are still sought after by scholars and students but could not be reprinted economically using traditional technology. The Cambridge Library Collection extends this activity to a wider range of books which are still of importance to researchers and professionals, either for the source material they contain, or as landmarks in the history of their academic discipline.

Drawing from the world-renowned collections in the Cambridge University Library, and guided by the advice of experts in each subject area, Cambridge University Press is using state-of-the-art scanning machines in its own Printing House to capture the content of each book selected for inclusion. The files are processed to give a consistently clear, crisp image, and the books finished to the high quality standard for which the Press is recognised around the world. The latest print-on-demand technology ensures that the books will remain available indefinitely, and that orders for single or multiple copies can quickly be supplied.

The Cambridge Library Collection will bring back to life books of enduring scholarly value (including out-of-copyright works originally issued by other publishers) across a wide range of disciplines in the humanities and social sciences and in science and technology.

A Series of Discourses on the Christian Revelation

Viewed in Connection with the Modern Astronomy

Thomas Chalmers

CAMBRIDGE UNIVERSITY PRESS

Cambridge, New York, Melbourne, Madrid, Cape Town, Singapore,
São Paolo, Delhi, Dubai, Tokyo

Published in the United States of America by Cambridge University Press, New York

www.cambridge.org
Information on this title: www.cambridge.org/9781108005272

This edition first published 1817
This digitally printed version 2009

ISBN 978-1-108-00527-2 Paperback

A SERIES OF DISCOURSES ON THE CHRISTIAN REVELATION, VIEWED IN CONNECTION WITH THE MODERN ASTRONOMY.

BY

THOMAS CHALMERS, D. D.

MINISTER OF THE TRON CHURCH, GLASGOW.

GLASGOW:

PRINTED BY JAMES HEDDERWICK,

FOR JOHN SMITH AND SON, GLASGOW;

WILLIAM WHYTE, EDINBURGH;

LONGMAN, HURST, REES, ORME, AND BROWN,

JOHN HATCHARD, GALE AND FENNER, THOMAS HAMILTON,

AND OGLES, DUNCAN, AND COCHRAN,

LONDON.

1817.

PREFACE.

The astronomical objection against the truth of the Gospel, does not occupy a very prominent place in any of our Treatises of Infidelity. It is often, however, met with in conversation—and we have known it to be the cause of serious perplexity and alarm in minds anxious for the solid establishment of their religious faith.

There is an imposing splendour in the science of astronomy; and it is not to be wondered at, if the light it throws, or appears to throw, over other tracks of speculation than those which are properly its own, should at times dazzle and mislead an inquirer. On this account, we think it were a service to what we deem a true and a righteous cause, could we succeed in dissipating this illusion; and in stripping Infidelity of those pretensions to enlargement, and to a certain air of philosophical greatness, by which it has often become so destructively alluring to the young, and the ardent, and the ambitious.

In my first Discourse, I have attempted a sketch of the Modern Astronomy—nor have I wished to throw any disguise over that comparative littleness which belongs to our planet, and which gives to the argument of Freethinkers all its plausibility.

This argument involves in it an assertion and an inference. The assertion is, that Christianity is a religion which professes to be designed for the single benefit of our world; and the inference is, that God cannot be the author of this religion, for he would not lavish on so insignificant a field, such peculiar and such distinguishing attentions, as are ascribed to him in the Old and New Testament.

Christianity makes no such profession. That it is designed for the single benefit of our world, is altogether a presumption of the Infidel himself—and feeling that this is not the only example of temerity which can be charged on the enemies of our faith, I have allotted my second Discourse to the attempt of demonstrating the utter repugnance of such a spirit with the cautious and enlightened philosophy of modern times.

In the course of this Sermon I have offered a tribute of acknowledgement to the theology of Sir Isaac Newton; and in such terms, as if not farther explained, may be liable to misconstruction. The grand circumstance of applause in the character of this great man, is, that unseduced by all the magnificence of his own discoveries, he had a solidity of mind which could resist their fascination; and keep him in steady attachment to that book, whose general evidences stamped upon it the impress of a real communication from heaven. This was the sole attribute of his theology which I had in my eye when I presumed to eulogise it. I do not think, that, amid the distraction and the engrossment of his other pursuits, he has at all times succeeded in his interpretation of the book; else he would never, in my apprehension, have abetted the leading doctrine of a sect or a system, which has now nearly dwindled away from public observation.

In my third Discourse I am silent as to the assertion, and attempt to combat the inference that is founded on it. I insist, that upon all the analogies of nature and of providence, we can lay no limit on the condescension of God, or on the multiplicity of his regards even to the very humblest departments of creation; and that it is not for us, who see the evidences of divine wisdom and care spread in such exhaustless profusion around us, to say, that the Deity would not lavish all the wealth of his wondrous attributes on the salvation even of our solitary species.

At this point of the argument I trust that the intelligent reader may be enabled to perceive in the adversaries of the gospel, a twofold dereliction from the maxims of the Baconian philosophy: that, in the first instance, the assertion which forms the groundwork of their argument, is gratuitously fetched out of an unknown region where they are utterly abandoned by the light of experience; and that, in the second instance, the inference they urge from it, is in the face of manifold and undeniable truths, all lying within the safe and accessible field of human observation.

In my subsequent Discourses, I proceed to the informations of the record. The Infidel objection, drawn from astronomy, may be considered as by this time disposed of; and if we have succeeded in clearing it away, so as to deliver the Christian testimony from all discredit upon this ground, then may we submit, on the strength of other evidences, to be guided by its information. We shall thus learn, that Christianity has a far more extensive bearing on the other orders of creation, than the Infidel is disposed to allow; and, whether he will own the authority of this information or not, he will, at least, be forced to admit, that the subject matter of the Bible itself is not chargeable with that objection which he has attempted to fasten upon it.

Thus, had my only object been the refutation of the Infidel argument, I might have spared the last Discourses of the Volume altogether. But the tracks of Scriptural information to which they directed me, I considered as worthy of prosecution on their own account—and I do think, that much may be gathered from these less observed portions of the field of revelation, to cheer, and to elevate, and to guide the believer.

But, in the management of such a discussion as this, though, for a great degree of this effect, it would require to be conducted in a far higher style than I am able to sustain, the taste of the human mind may be regaled; and its understanding put into a state of the most agreeable exercise. Now, this is quite distinct from the conscience being made to feel the force of a personal application; nor could I either bring this argument to its close in the pulpit, or offer it to the general notice of the world, without adverting, in the last Discourse, to a delusion, which, I fear, is carrying forward thousands, and tens of thousands, to an undone eternity.

I have closed the Volume with an Appendix of Scriptural Authorities. I found that I could not easily interweave them in the texture of the Work, and have, therefore, thought fit to present them in a separate form. I look for a twofold benefit from this exhibition—first, on those more general readers, who are ignorant of the Scriptures, and of the richness and variety which abound in them—and, secondly, on those narrow and intolerant professors, who take an alarm at the very sound and semblance of philosophy; and feel as if there was an utter irreconcileable antipathy between its lessons on the one hand, and the soundness and piety of the Bible on the other. It were well, I conceive, for our cause, that the latter could become a little more indulgent on this subject; that they gave up a portion of those ancient and hereditary prepossessions, which go so far to cramp and to enthral them; that they would suffer theology to take that wide range of argument and of illustration which belongs to her; and that, less sensitively jealous of any desecration being brought upon the Sabbath, or the pulpit, they would suffer her freely to announce all those truths, which either serve to protect Christianity from the contempt of science; or to protect the teachers of Christianity from those invasions, which are practised both on the sacredness of the office, and on the solitude of its devotional and intellectual labours.

I shall only add, for the information of readers at a distance, that these Discourses were chiefly delivered on the occasion of the week-day sermon that is preached in rotation by the Ministers of Glasgow.

CONTENTS.

DISCOURSE I.

DISCOURSE II.

DISCOURSE III.

ON THE EXTENT OF THE DIVINE CONDESCENSION.

DISCOURSE IV.

ON THE KNOWLEDGE OF MAN'S MORAL HISTORY IN THE DISTANT PLACES OF CREATION.

DISCOURSE V.

ON THE SYMPATHY THAT IS FELT FOR MAN IN THE DISTANT PLACES OF CREATION.

DISCOURSE VI.

DISCOURSE VII.

DISCOURSE I.

A SKETCH OF THE MODERN ASTRONOMY.

"When I consider thy heavens, the work of thy fingers, the moon and the stars, which thou hast ordained; What is man, that thou art mindful of him? and the son of man, that thou visitest him?"—PSALM viii. 3, 4.

IN the reasonings of the Apostle Paul, we cannot fail to observe, how studiously he accommodates his arguments to the pursuits or principles or prejudices of the people whom he was addressing. He often made a favourite opinion of their own the starting point of his explanation; and educing a dexterous but irresistible train of argument from some principle upon which each of the parties had a common understanding, did he force them out of all their opposition, by a weapon of their own choosing—

nor did he scruple to avail himself of a Jewish peculiarity, or a heathen superstition, or a quotation from Greek poetry, by which he might gain the attention of those whom he laboured to convince, and by the skilful application of which, he might "shut them up unto the faith."

Now, when Paul was thus addressing one class of an assembly, or congregation, another class might, for the time, have been shut out of all direct benefit and application from his arguments. When he wrote an Epistle to a mixed assembly of Christianised Jews and Gentiles, he had often to direct such a process of argument to the former, as the latter would neither require nor comprehend. Now, what should have been the conduct of the Gentiles at the reading of that part of the Epistle which bore almost an exclusive reference to the Jews? Should it be impatience at the hearing of something for which they had no relish or understanding? Should it be a fretful disappointment, because every thing that was said, was not said for their edification? Should it be angry discontent with the Apostle, because, leaving them in the dark, he had brought forward nothing for them, through the whole extent of so many successive chapters? Some of them may have felt in this way; but surely it would have been vastly more Christian to have sat with meek and unfeigned patience, and to have rejoiced that the great Apostle had undertaken the management of those obstinate prejudices, which kept back so many human beings from the participation of the Gospel. And should Paul have had reason to rejoice, that, by the success of his arguments, he had reconciled one or any number of Jews to Christianity, then it was the part of these Gentiles, though receiving no direct or personal benefit from the arguments, to have blessed God, and rejoiced along with him.

Conceive that Paul were at this moment alive, and zealously engaged in the work of pressing the Christian religion on the acceptance of the various classes of society. Should he not still have acted on the principle of being all things to all men? Should he not have accommodated his discussion to the prevailing taste, and literature, and philosophy of the times? Should he not have closed with the people, whom he was addressing, on some favourite principle of their own; and, in the prosecution of this principle, might he not have got completely beyond the comprehension of a numerous class of zealous, humble, and devoted Christians? Now, the question is not, how these would conduct themselves in such circumstances? but, how should they do it? Would it be right in them to sit with impatience, because the argument of the Apostle contained in it nothing in the way of comfort or edification to themselves? Should not the benevolence of the Gospel give a different direction to their feelings? And, instead of that narrow, exclusive, and monopolising spirit, which I fear is too characteristic of the more declared professors of the truth as it is in Jesus, ought they not to be patient, and to rejoice; when to philosophers, and to men of literary accomplishment, and to those who have the direction of the public taste among the upper walks of society, such arguments are addressed as may bring home to their acceptance also, "the words of this life?" It is under the impulse of these considerations, that I have, with some hesitation, prevailed upon myself to attempt an argument which I think fitted to soften and subdue those prejudices which lie at the bottom of what may be called the infidelity of natural science; if possible to bring over to the humility of the Gospel, those who expatiate with delight on the wonders and the sublimities of creation; and to convince them that a loftier wisdom still than that even of their high and honourable acquirements, is the wisdom of him who is resolved to know nothing but Jesus Christ and him crucified.

It is truly a most Christian exercise, to extract a sentiment of piety from the works and the appearances of nature. It has the authority of the Sacred Writers upon its side, and even our Saviour himself gives it the weight and the solemnity of his example. "Behold the lilies of the field; they toil not, neither do they spin, yet your heavenly Father careth for them." He expatiates on the beauty of a single flower, and draws from it the delightful argument of confidence in God. He gives us to see that taste may be combined with piety, and that the same heart may be occupied with all that is serious in the contemplations of religion, and be at the same time alive to the charms and the loveliness of nature.

The Psalmist takes a still loftier flight. He leaves the world, and lifts his imagination to that mighty expanse which spreads above it and around it. He wings his way through space, and wanders in thought over its immeasurable regions. Instead of a dark and unpeopled solitude, he sees it crowded with splendour, and filled with the energy of the Divine presence. Creation rises in its immensity before him, and the world, with all which it inherits, shrinks into littleness at a contemplation so vast and so overpowering. He wonders that he is not overlooked amid the grandeur and the variety which are on every side of him, and passing upward from the majesty of nature to the majesty of nature's Architect, he exclaims, "What is man that thou art mindful of him, or the son of man that thou shouldest deign to visit him?"

It is not for us to say, whether inspiration revealed to the Psalmist the wonders of the modern astronomy. But even though the mind be a perfect stranger to the science of these enlightened times, the heavens present a great and an elevating spectacle, an immense concave reposing upon the circular boundary of the world, and the innumerable lights which are suspended from on high, moving with solemn regularity along its surface. It seems to have been at night that the piety of the Psalmist was awakened by this contemplation, when the moon and the stars were visible, and not when the sun had risen in his strength, and thrown a splendour around him, which bore down and eclipsed all the lesser glories of the firmament. And there is much in the scenery of a nocturnal sky, to lift the soul to pious contemplation. That moon, and these stars, what are they? They are detached from the world, and they lift you above it. You feel withdrawn from the earth, and rise in lofty abstraction above this little theatre of human passions and human anxieties. The mind abandons itself to reverie, and is transferred in the ecstacy of its thoughts, to distant and unexplored regions. It sees nature in the simplicity of her great elements, and it sees the God of nature invested with the high attributes of wisdom and majesty.

But what can these lights be? The curiosity of the human mind is insatiable, and the mechanism of these wonderful heavens, has, in all ages, been its subject and its employment. It has been reserved for these latter times, to resolve this great and interesting question. The sublimest powers of philosophy have been called to the exercise, and astronomy may now be looked upon as the most certain and best established of the sciences.

We all know that every visible object appears less in magnitude as it recedes from the eye. The lofty vessel as it retires from the coast, shrinks into littleness, and at last appears in the form of a small speck on the verge of the horizon. The eagle with its expanded wings, is a noble object; but when it takes its flight into the upper regions of the air, it becomes less to the eye, and is seen like a dark spot upon the vault of heaven. The same is true of all magnitude. The heavenly bodies appear small to the eye of an inhabitant of this earth, only from the immensity of their distance. When we talk of hundreds of millions of miles, it is not to be listened to as incredible. For remember that we are talking of those bodies which are scattered over the immensity of space, and that space knows no termination. The conception is great and difficult, but the truth is unquestionable. By a process of measurement which it is unnecessary at present to explain, we have ascertained first the distance, and then the magnitude of some of those bodies which roll in the firmament; that the sun which presents itself to the eye under so diminutive a form, is really a globe, exceeding, by many thousands of times, the dimensions of the earth which we inhabit; that the moon itself has the magnitude of a world; and that even a few of those stars, which appear like so many lucid points to the unassisted eye of the observer, expand into large circles upon the application of the telescope, and are some of them much larger than the ball which we tread upon, and to which we proudly apply the denomination of the universe.

Now, what is the fair and obvious presumption? The world in which we live, is a round ball of a determined magnitude, and occupies its own place in the firmament. But when we explore the unlimited tracts of that space, which is every where around us, we meet with other balls of equal or superior magnitude, and from which our earth would either be invisible, or appear as small as any of those twinkling stars which are seen on the canopy of heaven. Why then suppose that this little spot, little at least in the immensity which surrounds it, should be the exclusive abode of life and of intelligence? What reason to think that those mightier globes which roll in other parts of creation, and which we have discovered to be worlds in magnitude, are not also worlds in use and in dignity? Why should we think that the great Architect of nature, supreme in wisdom as he is in power, would call these stately mansions into existence and leave them unoccupied? When we cast our eye over the broad sea, and look at the country on the other side, we see nothing but the blue land stretching obscurely over the distant horizon. We are too far away to perceive the richness of its scenery, or to hear the sound of its population. Why not extend this principle to the still more distant parts of the universe? What though, from this remote point of observation, we can see nothing but the naked roundness of yon planetary orbs? Are we therefore to say, that they are so many vast and unpeopled solitudes; that desolation reigns in every part of the universe but ours; that the whole energy of the divine attributes is expended on one insignificant corner of these mighty works; and that to this earth alone, belongs the bloom of vegetation, or the blessedness of life, or the dignity of rational and immortal existence?

But this is not all. We have something more than the mere magnitude of the planets to allege, in favour of the idea that they are inhabited. We know that this earth turns round upon itself; and we observe that all those celestial bodies, which are accessible to such an observation, have the same movement. We know that the earth performs a yearly revolution round the sun; and we can detect in all the planets which compose our system, a revolution of the same kind, and under the same circumstances. They have the same succession of day and night. They have the same agreeable vicissitude of the seasons. To them, light and darkness succeed each other; and the gaiety of summer is followed by the dreariness of winter. To each of them the heavens present as varied and magnificent a spectacle; and this earth, the encompassing of which would require the labour of years from one of its puny inhabitants, is but one of the lesser lights which sparkle in their firmament. To them, as well as to us, has God divided the light from the darkness, and he has called the light day, and the darkness he has called night. He has said, let there be lights in the firmament of their heaven, to divide the day from the night; and let them be for signs, and for seasons, and for days, and for years; and let them be for lights in the firmament of heaven, to give lights upon their earth; and it was so. And God has also made to them great lights. To all of them he has given the sun to rule the day; and to many of them has he given moons to rule the night. To them he has made the stars also. And God has set them in the firmament of heaven, to give light unto their earth; and to rule over the day, and over the night, and to divide the light from the darkness; and God has seen that it was good.

In all these greater arrangements of divine wisdom, we can see that God has done the same things for the accommodation of the planets that he has done for the earth which we inhabit. And shall we say, that the resemblance stops here, because we are not in a situation to observe it? Shall we say, that this scene of magnificence has been called into being merely for the amusement of a few astronomers? Shall we measure the counsels of heaven by the narrow impotence of the human faculties? or conceive, that silence and solitude reign throughout the mighty empire of nature; that the greater part of creation is an empty parade; and that not a worshipper of the Divinity is to be found through the wide extent of yon vast and immeasurable regions?

It lends a delightful confirmation to the argument, when, from the growing perfection of our instruments, we can discover a new point of resemblance between our Earth and the other bodies of the planetary system. It is now ascertained, not merely that all of them have their day and night, and that all of them have their vicissitudes of seasons, and that some of them have their moons to rule their night and alleviate the darkness of it.— We can see of one, that its surface rises into inequalities, that it swells into mountains and stretches into valleys; of another, that it is surrounded by an atmosphere which may support the respiration of animals; of a third, that clouds are formed and suspended over it, which may minister to it all the bloom and luxuriance of vegetation; and of a fourth, that a white colour spreads over its northern regions, as its winter advances, and that on the approach of summer this whiteness is dissipated—giving room to suppose, that the element of water abounds in it, that it rises by evaporation into its atmosphere, that it freezes upon the application of cold, that it is precipitated in the form of snow, that it covers the ground with a fleecy mantle, which melts away from the heat of a more vertical sun; and that other worlds bear a resemblance to our own, in the same yearly round of beneficent and interesting changes.

Who shall assign a limit to the discoveries of future ages? Who can prescribe to science her boundaries, or restrain the active and insatiable curiosity of man within the circle of his present acquirements? We may guess with plausibility what we cannot anticipate with confidence. The day may yet be coming, when our instruments of observation shall be inconceivably more powerful. They may ascertain still more decisive points of resemblance. They may resolve the same question by the evidence of sense which is now so abundantly convincing by the evidence of analogy. They may lay open to us the unquestionable vestiges of art, and industry, and intelligence. We may see summer throwing its green mantle over these mighty tracts, and we may see them left naked and colourless after the flush of vegetation has disappeared. In the progress of years, or of centuries, we may trace the hand of cultivation spreading a new aspect over some portion of a planetary surface. Perhaps some large city, the metropolis of a mighty empire, may expand into a visible spot by the powers of some future telescope. Perhaps the glass of some observer, in a distant age, may enable him to construct the map of another world, and to lay down the surface of it in all its minute and topical varieties. But there is no end of conjecture, and to the men of other times we leave the full assurance of what we can assert with the highest probability, that yon planetary orbs are so many worlds, that they teem with life, and that the mighty Being who presides in high authority over this scene of grandeur and astonishment, has there planted the worshippers of his glory.

Did the discoveries of science stop here, we have enough to justify the exclamation of the Psalmist, " What is man that thou art mindful of him, or the son of man that thou shouldest deign to visit him?" They widen the empire of creation far beyond the limits which were formerly assigned to it. They give us to see that yon sun, throned in the centre of his planetary system, gives light, and warmth, and the vicissitude of seasons, to an extent of surface, several hundreds of times greater than that of the earth which we inhabit. They lay open to us a number of worlds, rolling in their respective circles around this vast luminary—and prove, that the ball which we tread upon, with all its mighty burden of oceans and continents, instead of being distinguished from the others, is among the least of them; and, from some of the more distant planets, would not occupy a visible point in the concave of their firmament. They let us know, that though this mighty earth, with all its myriads of people, were to sink into annihilation, there are some worlds where an event so awful to us would be unnoticed and unknown, and others where it would be nothing more than the disappearance of a little star which had ceased from its twinkling. We should feel a sentiment of modesty at this just but humiliating representation. We should learn not to look on our earth as the universe of God, but one paltry and insignificant portion of it; that it is only one of the many mansions which the Supreme Being has created for the accommodation of his worshippers, and only one of the many worlds rolling in that flood of light which the sun pours around him to the outer limits of the planetary system.

But is there nothing beyond these limits? The planetary system has its boundary, but space has none; and if we wing our fancy there, do we only travel through dark and unoccupied regions? There are only five, or at most six, of the planetary orbs visible to the naked eye. What, then, is that multitude of other lights which sparkle in our firmament, and fill the whole concave of heaven with innumerable splendours? The planets are all attached to the sun; and, in circling around him, they do homage to that influence which binds them to perpetual attendance on this great luminary. But the other stars do not own his dominion. They do not circle around him. To all common observation, they remain immoveable; and each, like the independent sovereign of his own territory, appears to occupy the same inflexible position in the regions of immensity. What can we make of them? Shall we take our adventurous flight to explore these dark and untravelled dominions? What mean these innumerable fires lighted up in distant parts of the universe? Are they only made to shed a feeble glimmering over this little spot in the kingdom of nature? or do they serve a purpose worthier of themselves, to light up other worlds, and give animation to other systems?

The first thing which strikes a scientific observer of the fixed stars, is their immeasurable distance. If the whole planetary system were lighted up into a globe of fire, it would exceed, by many millions of times, the magnitude of this world, and yet only appear a small lucid point from the nearest of them. If a body were projected from the sun with the velocity of a cannon-ball, it would take hundreds of thousands of years before it described that mighty interval which separates the nearest of the fixed stars from our sun and from our system. If this earth, which moves at more than the inconceivable velocity of a million and a half miles a-day, were to be hurried from its orbit, and to take the same rapid flight over this immense tract, it would not have arrived at the termination of its journey, after taking all the time which has elapsed since the creation of the world. These are great numbers, and great calculations, and the mind feels its own impotency in attempting to grasp them. We can state them in words. We can exhibit them in figures. We can demonstrate them by the powers of a most rigid and infallible geometry. But no human fancy can summon up a lively or an adequate conception—can roam in its ideal flight over this immeasurable largeness—can take in this mighty space in all its grandeur, and in all its immensity—can sweep the outer boundaries of such a creation—or lift itself up to the majesty of that great and invisible arm, on which all is suspended.

But what can those stars be which are seated so far beyond the limits of our planetary system? They must be masses of immense magnitude, or they could not be seen at the distance of place which they occupy. The light which they give must proceed from themselves, for the feeble reflection of light from some other quarter, would not carry through such mighty tracts to the eye of an observer. A body may be visible in two ways. It may be visible from its own light as the flame of a candle, or the brightness of a fire, or the brilliancy of yonder glorious sun which lightens all below, and is the lamp of the world. Or it may be visible from the light which falls upon it, as the body which receives its light from the taper that falls upon it—or the whole assemblage of objects on the surface of the earth, which appear only when the light of day rests upon them—or the moon, which, in that part of it that is towards the sun, gives out a silvery whiteness to the eye of the observer, while the other part forms a black and invisible space in the firmament—or as the planets, which shine only because the sun shines upon them, and which, each of them, present the appearance of a dark spot on the side that is turned away from it. Now apply this question to the fixed stars. Are they luminous of themselves, or do they derive their light from the sun, like the bodies of our planetary system? Think of their immense distance, and the solution of this question becomes evident. The sun, like any other body, must dwindle into a less apparent magnitude as you retire from it. At the prodigious distance even of the very nearest of the fixed stars, it must have shrunk into a small indivisible point. In short, it must have become a star itself, and could shed no more light than a single individual of those glimmering myriads, the whole assemblage of which cannot dissipate and can scarcely alleviate the midnight darkness of our world. These stars are visible to us, not because the sun shines upon them, but because they shine of themselves, because they are so many luminous bodies scattered over the tracts of immensity—in a word, because they are so many suns, each throned in the centre of his own dominions, and pouring a flood of light over his own portion of these unlimitable regions.

At such an immense distance for observation, it is not to be supposed, that we can collect many points of resemblance between the fixed stars, and the solar star which forms the centre of our planetary system. There is one point of resemblance, however, which has not escaped the penetration of our astronomers. We know that our sun turns round upon himself, in a regular period of time. We also know, that there are dark spots scattered over his surface, which, though invisible to the naked eye, are perfectly noticeable by our instruments. If these spots existed in greater quantity upon one side than upon another, it would have the general effect of making that side darker, and the revolution of the sun must, in such a case, give us a brighter and a fainter side, by regular alternations. Now, there are some of the fixed stars which present this appearance. They present us with periodical variations of light. From the splendour of a star of the first or second magnitude, they fade away into some of the inferior magnitudes—and one, by becoming invisible, might give reason to apprehend that we had lost him altogether—but we can still recognise him by the telescope, till at length he reappears in his own place, and, after a regular lapse of so many days and hours, recovers his original brightness. Now, the fair inference from this, is, that the fixed stars, as they resemble our sun in being so many luminous masses of immense magnitude, they resemble him in this also, that each of them turns round upon his own axis; so that if any of them should have an inequality in the brightness of their sides, this revolution is rendered evident, by the regular variations in the degree of light which it undergoes.

Shall we say, then, of these vast luminaries, that they were created in vain? Were they called into existence for no other purpose than to throw a tide of useless splendour over the solitudes of immensity? Our sun is only one of these luminaries, and we know that he has worlds in his train. Why should we strip the rest of this princely attendance? Why may not each of them be the centre of his own system, and give light to his own worlds? It is true that we see them not, but could the eye of man take its flight into those distant regions, it should lose sight of our little world, before it reached the outer limits of our system—the greater planets should disappear in their turn—before it had described a small portion of that abyss which separates us from the fixed stars, the sun should decline into a little spot, and all its splendid retinue of worlds be lost in the obscurity of distance—he should, at last, shrink into a small indivisible atom, and all that could be seen of this magnificent system, should be reduced to the glimmering of a little star. Why resist any longer the grand and interesting conclusion? Each of these stars may be the token of a system as vast and as splendid as the one which we inhabit. Worlds roll in these distant regions; and these worlds must be the mansions of life and of intelligence. In yon gilded canopy of heaven, we see the broad aspect of the universe, where each shining point presents us with a sun, and each sun with a system of worlds—where the Divinity reigns in all the grandeur of his attributes—where he peoples immensity with his wonders; and travels in the greatness of his strength through the dominions of one vast and unlimited monarchy.

The contemplation has no limits. If we ask the number of suns and of systems, the unassisted eye of man can take in a thousand, and the best telescope which the genius of man has constructed can take in eighty millions. But why subject the dominions of the universe to the eye of man, or to the powers of his genius? Fancy may take its flight far beyond the ken of eye or of telescope. It may expatiate in the outer regions of all that is visible— and shall we have the boldness to say, that there is nothing there? that the wonders of the Almighty are at an end, because we can no longer trace his footsteps? that his omnipotence is exhausted, because human art can no longer follow him? that the creative energy of God has sunk into repose, because the imagination is enfeebled by the magnitude of its efforts, and can keep no longer on the wing through those mighty tracts, which shoot far beyond what eye hath seen, or the heart of man hath conceived— which sweep endlessly along, and merge into an awful and mysterious infinity?

Before bringing to a close this rapid and imperfect sketch of our modern astronomy, it may be right to advert to two points of interesting speculation, both of which serve to magnify our conceptions of the universe, and, of course, to give us a more affecting sense of the comparative insignificance of this our world. The first is suggested by the consideration, that, if a body be struck in the direction of its centre, it obtains, from this course, a progressive motion, but without any movement of revolution being at the same time impressed upon it. It simply goes forward, but does not turn round upon itself. But, again, should the stroke not be in the direction of the centre—should the line which joins the point of percussion to the centre, make an angle with that line in which the impulse was communicated, then the body is both made to go forward in space, and also to wheel upon its axis. In this way, each of our planets may have had their compound motion communicated to it by one single impulse; and, on the other hand, if ever the rotatory motion be communicated by one blow, then the progressive motion must go along with it. In order to have the first motion without the second, there must be a twofold force applied to the body in opposite directions. It must be set agoing in the same way as a spinning-top, so as to revolve about an axis, and to keep unchanged its situation in space. The planets have both motions; and, therefore, may have received them by one and the same impulse. The sun, we are certain, has one of these motions. He has a movement of revolution. If spun round his axis by two opposite forces, one on each side of him, he may have this movement, and retain an inflexible position in space. But, if this movement was given him by one stroke, he must have a progressive motion, along with a whirling motion; or, in other words, he is moving forward; he is describing a tract in space; and, in so doing, he carries all his planets and all their secondaries along with him.

But, at this stage of the argument, the matter only remains a conjectural point of speculation. The sun may have had his rotation impressed upon him by a spinning impulse; or, without recurring to secondary causes at all, this movement may be coeval with his being, and he may have derived both the one and the other from an immediate fiat of the Creator. But, there is an actually observed phenomenon of the heavens, which advances the conjecture into a probability. In the course of ages, the stars in one quarter of the celestial sphere are apparently receding from each other; and, in the opposite quarter, they are apparently drawing nearer to each other. If the sun be approaching the former quarter, and receding from the latter, this phenomenon admits of an easy explanation, and we are furnished with a magnificent step in the scale of the Creator's workmanship. In the same manner as the planets, with their satellites, revolve round the sun, may the sun, with all his tributaries, be moving, in common with other stars, around some distant centre, from which there emanates an influence to bind and to subordinate them all. They may be kept from approaching each other, by a centrifugal force; without which, the laws of attraction might consolidate, into one stupendous mass, all the distinct globes, of which the universe is composed. Our sun may, therefore, be only one member of a higher family—taking his part, along with millions of others, in some loftier system of mechanism, by which they are all subjected to one law, and to one arrangement—describing the sweep of such an orbit in space, and completing the mighty revolution in such a period of time, as to reduce our planetary seasons, and our planetary movements, to a very humble and fractionary rank in the scale of a higher astronomy. There is room for all this in immensity; and there is even argument for all this, in the records of actual observation; and, from the whole of this speculation, do we gather a new emphasis to the lesson, how minute is the place, and how secondary is the importance of our world, amid the glories of such a surrounding magnificence.

But, there is still another very interesting tract of speculation, which has been opened up to us by the more recent observations of astronomy. What we allude to, is the discovery of the *nebulæ*. We allow that it is but a dim and indistinct light which this discovery has thrown upon the structure of the universe; but still it has spread before the eye of the mind a field of very wide and lofty contemplation. Anterior to this discovery, the universe might appear to have been composed of an indefinite number of suns, about equi-distant from each other, uniformly scattered over space, and each encompassed by such a planetary attendance as takes place in our own system. But, we have now reason to think, that, instead of lying uniformly, and in a state of equi-distance from each other, they are arranged into distinct clusters—that, in the same manner as the distance of the nearest fixed stars so inconceivably superior to that of our planets from each other, marks the separation of the solar systems, so the distance of two contiguous clusters may be so inconceivably superior to the reciprocal distance of those fixed stars which belong to the same cluster, as to mark an equally distinct separation of the clusters, and to constitute each of them an individual member of some higher and more extended arrangement. This carries us upwards through another ascending step in the scale of magnificence, and there leaves us wildering in the uncertainty, whether even here the wonderful progression is ended; and, at all events, fixes the assured conclusion in our minds, that, to an eye which could spread itself over the whole, the mansion which accommodates our species might be so very small as to lie wrapped in microscopical concealment; and, in reference to the only Being who possesses this universal eye, well might we say, "What is man that thou art mindful of him, or the son of man that thou shouldest deign to visit him?"

And, after all, though it be a mighty and difficult conception, yet who can question it? What is seen may be nothing to what is unseen; for what is seen is limited by the range of our instruments. What is unseen has no limit; and, though all which the eye of man can take in, or his fancy can grasp at, were swept away, there might still remain as ample a field, over which the Divinity may expatiate, and which he may have peopled with innumerable worlds. If the whole visible creation were to disappear, it would leave a solitude behind it—but to the Infinite Mind, that can take in the whole system of nature, this solitude might be nothing; a small unoccupied point in that immensity which surrounds it, and which he may have filled with the wonders of his omnipotence. Though this earth were to be burned up, though the trumpet of its dissolution were sounded, though yon sky were to pass away as a scroll, and every visible glory, which the finger of the Divinity has inscribed on it, were to be put out for ever—an event, so awful to us, and to every world in our vicinity, by which so many suns would be extinguished, and so many varied scenes of life and of population would rush into forgetfulness—what is it in the high scale of the Almighty's workmanship? a mere shred, which, though scattered into nothing, would leave the universe of God one entire scene of greatness and of majesty. Though this earth, and these heavens, were to disappear, there are other worlds, which roll afar; the light of other suns shines upon them; and the sky which mantles them, is garnished with other stars. Is it presumption to say, that the moral world extends to these distant and unknown regions? that they are occupied with people? that the charities of home and of neighbourhood flourish there? that the praises of God are there lifted up, and his goodness rejoiced in? that piety has its temples and its offerings? and the richness of the divine attributes is there felt and admired by intelligent worshippers?

And what is this world in the immensity which teems with them—and what are they who occupy it? The universe at large would suffer as little, in its splendour and variety, by the destruction of our planet, as the verdure and sublime magnitude of a forest would suffer by the fall of a single leaf. The leaf quivers on the branch which supports it. It lies at the mercy of the slightest accident. A breath of wind tears it from its stem, and it lights on the stream of water which passes underneath. In a moment of time, the life, which we know, by the microscope, it teems with, is extinguished; and, an occurrence so insignificant in the eye of man, and on the scale of his observation, carries in it, to the myriads which people this little leaf, an event as terrible and as decisive as the destruction of a world. Now, on the grand scale of the universe, we, the occupiers of this ball, which performs its little round among the suns and the systems that astronomy has unfolded—we may feel the same littleness, and the same insecurity. We differ from the leaf only in this circumstance, that it would require the operation of greater elements to destroy us. But these elements exist. The fire which rages within, may lift its devouring energy to the surface of our planet, and transform it into one wide and wasting volcano. The sudden formation of elastic matter in the bowels of the earth—and it lies within the agency of known substances to accomplish this—may explode it into fragments. The exhalation of noxious air from below, may impart a virulence to the air that is around us; it may affect the delicate proportion of its ingredients; and the whole of animated nature may wither and die under the malignity of a tainted atmosphere. A blazing comet may cross this fated planet in its orbit, and realise all the terrors which superstition has conceived of it. We cannot anticipate with precision the consequences of an event which every astronomer must know to lie within the limits of chance and probability. It may hurry our globe towards the sun—or drag it to the outer regions of the planetary system—or give it a new axis of revolution—and the effect, which I shall simply announce, without explaining it, would be to change the place of the ocean, and bring another mighty flood upon our islands and continents. These are changes which may happen in a single instant of time, and against which nothing known in the present system of things provides us with any security. They might not annihilate the earth, but they would unpeople it; and we who tread its surface with such firm and assured footsteps, are at the mercy of devouring elements, which, if let loose upon us by the hand of the Almighty, would spread solitude, and silence, and death, over the dominions of the world.

Now, it is this littleness, and this insecurity, which make the protection of the Almighty so dear to us, and bring, with such emphasis, to every pious bosom, the holy lessons of humility and gratitude. The God who sitteth above, and presides in high authority over all worlds, is mindful of man; and, though at this moment his energy is felt in the remotest provinces of creation, we may feel the same security in his providence, as if we were the objects of his undivided care. It is not for us to bring our minds up to this mysterious agency. But, such is the incomprehensible fact, that the same Being, whose eye is abroad over the whole universe, gives vegetation to every blade of grass, and motion to every particle of blood which circulates through the veins of the minutest animal; that, though his mind takes into its comprehensive grasp, immensity and all its wonders, I am as much known to him as if I were the single object of his attention; that he marks all my thoughts; that he gives birth to every feeling and every movement within me; and that, with an exercise of power which I can neither describe nor comprehend, the same God who sits in the highest heaven, and reigns over the glories of the firmament, is at my right hand, to give me every breath which I draw, and every comfort which I enjoy.

But this very reflection has been appropriated to the use of Infidelity, and the very language of the text has been made to bear an application of hostility to the faith. "What is man, that

God should be mindful of him, or the son of man, that he should deign to visit him?" Is it likely, says the Infidel, that God would send his eternal Son, to die for the puny occupiers of so insignificant a province in the mighty field of his creation? Are we the befitting objects of so great and so signal an interposition? Does not the largeness of that field which astronomy lays open to the view of modern science, throw a suspicion over the truth of the gospel history; and how shall we reconcile the greatness of that wonderful movement which was made in heaven for the redemption of fallen man, with the comparative meanness and obscurity of our species?

This is a popular argument against Christianity, not much dwelt upon in books, but we believe, a good deal insinuated in conversation, and having no small influence on the amateurs of a superficial philosophy. At all events, it is right that every such argument should be met, and manfully confronted; nor do we know a more discreditable surrender of our religion, than to act as if she had any thing to fear from the ingenuity of her most accomplished adversaries.

The author of the following treatise, engages in his present undertaking, under the full impression, that a something may be found with which to combat Infidelity in all its forms; that the truth of God and of his message, admits of a noble and decisive manifestation, through every mist which the pride, or the prejudice, or the sophistry of man may throw around it; and elevated as the wisdom of him may be, who has ascended the heights of science, and poured the light of demonstration over the most wondrous of nature's mysteries, that even out of his own principles, it may be proved how much more elevated is the wisdom of him who sits with the docility of a little child, to his Bible, and casts down to its authority, all his lofty imaginations.

DISCOURSE II.

THE MODESTY OF TRUE SCIENCE.

"And if any man think that he knoweth any thing, he knoweth nothing yet as he ought to know."—1 Cor. viii. 2.

THERE is much profound and important wisdom in that proverb of Solomon, where it is said, that the heart knoweth its own bitterness. It forms part of a truth still more comprehensive, that every man knoweth his own peculiar feelings, and difficulties, and trials, far better than he can get any of his neighbours to perceive them. It is natural to us all, that we should desire to engross, to the uttermost, the sympathy of others with what is most painful to the sensibilities of our own bosom, and with what is most aggravating in the hardships of our own situation. But, labour it as we may, we cannot, with every power of expression, make an adequate conveyance, as it were, of all our sensations, and of all our circumstances, into another understanding. There is a something in the intimacy of a man's own experience, which he cannot make to pass entire into the heart and mind even of his most familiar companion—and thus it is, that he is so often defeated in his attempts to obtain a full and a cordial possession of his sympathy. He is mortified, and he wonders at the obtuseness of the people around him—and how he cannot get them to enter into the justness of his complainings—nor to feel the point upon which turn the truth and the reason of his remonstrances—nor to give their interested attention to the case of his peculiarities and of his wrongs—nor to kindle, in generous resentment, along with him, when he starts the topic of his indignation. He does not reflect, all the while, that, with every human being he addresses, there is an inner man, which forms a theatre of passions, and of interests, as busy, as crowded, and as fitted as his own to engross the anxious and the exercised feelings of a heart, which can alone understand its own bitterness, and lay a correct estimate on the burden of its own visitations. Every man we meet, carries about with him, in the unperceived solitude of his bosom, a little world of his own—and we are just as blind, and as insensible, and as dull, both of perception and of sympathy, about his engrossing objects, as he is about ours; and, did we suffer this observation to have all its weight upon us, it might serve to make us more candid, and more considerate of others. It might serve to abate the monopolizing selfishness of our nature. It might serve to soften down all the malignity which comes out of those envious contemplations that we are so apt to cast on the fancied ease and prosperity which are around us. It might serve to reconcile every man to his own lot, and dispose him to bear, with thankfulness, his own burden; and sure I am, if this train of sentiment were prosecuted with firmness, and calmness, and impartiality, it would lead to the conclusion, that each profession in life has its own peculiar pains, and its own besetting inconveniences—that, from the very bottom of society, up to the golden pinnacle which blazons upon its summit, there is much in the shape of care and of suffering to be found—that, throughout all the conceivable varieties of human condition, there are trials, which can neither be adequately told on the one side, nor fully understood on the other—that the ways of God to man are as equal in this, as in every department of his administration—and that, go to whatever quarter of human experience we may, we shall find how he has provided enough to exercise the patience, and to accomplish the purposes of a wise and a salutary discipline upon all his children.

I have brought forward this observation, that it may prepare the way for a second. There are perhaps no two sets of human beings, who comprehend less the movements, and enter less into the cares and concerns, of each other, than the wide and busy public on the one hand, and, on the other, those men of close and studious retirement, whom the world never hears of, save when, from their thoughtful solitude, there issues forth some splendid discovery, to set the world on a gaze of admiration. Then will the brilliancy of a superior genius draw every eye towards it—and the homage paid to intellectual superiority, will place its idol on a loftier eminence than all wealth or than all titles can bestow—and the name of the successful philosopher will circulate, in his own age, over the whole extent of civilised society, and be borne down to posterity in the characters of ever-during remembrance—and thus it is, that, when we look back on the days of Newton, we annex a kind of mysterious greatness to him, who, by the pure force of his understanding, rose to such a gigantic elevation above the level of ordinary men—and the kings and warriors of other days sink into insignificance around him—and he, at this moment, stands forth to the public eye, in a prouder array of glory than circles the memory of all the men of former generations—and, while all the vulgar grandeur of other days is now mouldering in forgetfulness, the achievements of our great astronomer are still fresh in the veneration of his countrymen, and they carry him forward on the stream of time, with a reputation ever gathering, and the triumphs of a distinction that will never die.

Now, the point that I want to impress upon you is, that the same public, who are so dazzled and overborne by the lustre of all this superiority, are utterly in the dark as to what that is which confers its chief merit on the philosophy of Newton. They see the result of his labours, but they know not how to appreciate the difficulty or the extent of them. They look on the stately edifice he has reared, but they know not what he had to do in settling the foundation which gives to it all its stability—nor are they aware what painful encounters he had to make, both with the natural predilections of his own heart, and with the prejudices of others, when employed on the work of laying together its unperishing materials. They have never heard of the controversies which this man, of peaceful unambitious modesty, had to sustain, with all that was proud, and all that was intolerant in the philosophy of the age. They have never, in thought, entered that closet which was the scene of his patient and profound exercises—nor have they gone along with him, as he gave his silent hours to the labours of the midnight oil, and plied that unwearied task, to which the charm of lofty contemplation had allured him—nor have they accompanied him through all the workings of that wonderful mind, from which, as from the recesses of a laboratory, there came forth such gleams and processes of thought as shed an effulgency over the whole amplitude of nature. All this, the public have not done; for of this the great majority, even of the reading and cultivated public, are utterly incapable; and, therefore, is it, that they need to be told what that is, in which the main distinction of his philosophy lies; that, when labouring in other fields of investigation, they may know how to borrow from his safe example, and how to profit by that superior wisdom which marked the whole conduct of his understanding.

Let it be understood, then, that they are the positive discoveries of Newton, which, in the eye of a superficial public, confer upon him all his reputation. He discovered the mechanism of the planetary system. He discovered the composition of light. He discovered the cause of those alternate movements which take place on the waters of the ocean. These form his actual and his visible achievements. These are what the world look at as the monuments of his greatness. These are doctrines by which he has enriched the field of philosophy; and thus it is, that the whole of his merit is supposed to lie in having had the sagacity to perceive, and the vigour to lay hold of the proofs, which conferred upon these doctrines all the establishment of a most rigid and conclusive demonstration.

But, while he gets all his credit, and all his admiration for those articles of science which he has added to the creed of philosophers, he deserves as much credit and admiration for those articles which he kept out of this creed, as for those which he introduced into it. It was the property of his mind, that it kept a tenacious hold of every one position which had proof to substantiate it—but it forms a property equally characteristic, and which, in fact, gives its leading peculiarity to the whole spirit and style of his investigations, that he put a most determined exclusion on every one position that was destitute of such proof. He would not admit the astronomical theories of those who went before him, because they had no proof. He would not give in to their notions about the planets wheeling their rounds in whirlpools of ether—for he did not see this ether—he had no proof of its existence—and, besides, even supposing it to exist, it would not have impressed, on the heavenly bodies, such movements as met his observation. He would not submit his judgement to the reigning systems of the day—for, though they had authority to recommend them, they had no proof; and thus it is, that he evinced the strength and the soundness of his philosophy, as much by his decisions upon those doctrines of science which he rejected, as by his demonstration of those doctrines of science which he was the first to propose, and which now stand out to the eye of posterity as the only monuments to the force and superiority of his understanding.

He wanted no other recommendation for any one article of science, than the recommendation of evidence—and, with this recommendation, he opened to it the chamber of his mind, though authority scowled upon it, and taste was disgusted by it, and fashion was ashamed of it, and all the beauteous speculation of former days was cruelly broken up by this new announcement of the better philosophy, and scattered like the fragments of an aerial vision, over which the past generations of the world had been slumbering their profound and their pleasing reverie. But, on the other hand, should the article of science want the recommendation of evidence, he shut against it all the avenues of his understanding—aye, and though all antiquity lent their suffrages to it, and all eloquence had thrown around it the most attractive brilliancy, and all habit had incorporated it with every system of every seminary in Europe, and all fancy had arrayed it in graces of the most tempting solicitation; yet was the steady and inflexible mind of Newton proof against this whole weight of authority and allurement, and, casting his cold and unwelcome look at the specious plausibility, he rebuked it from his presence. The strength of his philosophy lay as much in refusing admittance to that which wanted evidence, as in giving a place and an occupancy to that which possessed it. In that march of intellect, which led him onwards through the rich and magnificent field of his discoveries, he pondered every step; and, while he advanced with a firm and assured movement, wherever the light of evidence carried him, he never suffered any glare of imagination or of prejudice to seduce him from his path.

Sure I am, that, in the prosecution of his wonderful career, he found himself on a way beset with temptation upon every side of him. It was not merely that he had the reigning taste and philosophy of the times to contend with. But, he expatiated on a lofty region, where, in all the giddiness of success, he might have met with much to solicit his fancy, and tempt him to some devious speculation. Had he been like the majority of other men, he would have broken free from the fetters of a sober and chastised understanding, and, giving wing to his imagination, had done what philosophers have done after him—been carried away by some meteor of their own forming, or found their amusement in some of their own intellectual pictures, or palmed some loose and confident plausibilities of their own upon the world. But, Newton stood true to his principle, that he would take up with nothing which wanted evidence, and he kept by his demonstrations, and his measurements, and his proofs; and, if it be true that he who ruleth his own spirit is greater than he who taketh a city, there was won, in the solitude of his chamber, many a repeated victory over himself, which should give a brighter lustre to his name than all the conquests he has made on the field of discovery, or than all the splendour of his positive achievements.

I trust you understand, how, though it be one of the maxims of the true philosophy, never to shrink from a doctrine which has evidence on its side, it is another maxim, equally essential to it, never to harbour any doctrine when this evidence is wanting. Take these two maxims along with you, and you will be at no loss to explain the peculiarity, which, more than any other, goes both to characterise and to ennoble the philosophy of Newton. What I allude to, is, the precious combination of its strength and of its modesty. On the one hand, what greater evidence of strength than the fulfilment of that mighty enterprise, by which the heavens have been made its own, and the mechanism of unnumbered worlds has been brought within the grasp of the human understanding? Now, it was by walking in the light of sound and competent evidence, that all this was accomplished. It was by the patient, the strenuous, the unfaltering application of the legitimate instruments of discovery. It was by touching that which was tangible, and looking to that which was visible, and computing that which was measurable, and, in one word, by making a right and a reasonable use of all that proof which the field of nature around us has brought within the limit of sensible observation. This is the arena on which the modern philosophy has won all her victories, and fulfilled all her wondrous achievements, and reared all her proud and enduring monuments, and gathered all her magnificent trophies to that power of intellect with which the hand of a bounteous heaven has so richly gifted the constitution of our species.

But, on the other hand, go beyond the limits of sensible observation, and, from that moment, the genuine disciples of this enlightened school cast all their confidence and all their intrepidity away from them. Keep them on the firm ground of experiment, and none more bold and more decisive in their announcements of all that they have evidence for—but, off this ground, none more humble, or more cautious of any thing like positive announcements, than they. They choose neither to know, nor to believe, nor to assert, where evidence is a wanting, and they will sit, with all the patience of a scholar to his task, till they have found it. They are utter strangers to that haughty confidence with which some philosophers of the day sport the plausibilities of unauthorised speculation, and by which, unmindful of the limit that separates the region of sense from the region of conjecture, they make their blind and their impetuous inroads into a province which does not belong to them. There is no one object to which the exercised mind of a true Newtonian disciple is more familiarised than this limit, and it serves as a boundary by which he shapes, and bounds, and regulates, all the enterprises of his philosophy. All the space which lies within this limit, he cultivates to the uttermost, and it is by such successive labours, that every year which rolls over the world, is witnessing some new contribution to experimental science, and adding to the solidity and aggrandisement of this wonderful fabric. But, if true to their own principle, then, in reference to the forbidden ground which lies without this limit, those very men, who, on the field of warranted exertion, evinced all the hardihood and vigour of a full grown understanding, show, on every subject where the light of evidence is withheld from them, all the modesty of children. They give you positive opinion only when they have indisputable proof—but, when they have no such proof, then they have no such opinion. The single principle of their respect to truth, secures their homage for every one position, where the evidence of truth is present, and, at the same time, begets an entire diffidence about every one position, from which this evidence is disjoined. And thus you may understand, how the first man in the accomplishments of philosophy, which the world ever saw, sat at the book of nature in the humble attitude of its interpreter and its pupil—how all the docility of conscious ignorance threw a sweet and softening lustre around the radiance even of his most splendid discoveries—and, while the flippancy of a few superficial acquirements is enough to place a philosopher of the day on the pedestal of his fancied elevation, and to vest him with an assumed lordship over the whole domain of natural and revealed knowledge; I cannot forbear to do honour to the unpretending greatness of Newton, than whom I know not if there ever lighted on the face of our world, one in the character of whose admirable genius so much force and so much humility were more attractively blended.

I now propose to carry you forward, by a few simple illustrations, to the argument of this day. All the sublime truths of the modern astronomy lie within the field of actual observation, and have the firm evidence to rest upon of all that information which is conveyed to us by the avenue of the senses. Sir Isaac Newton never went beyond this field, without a reverential impression upon his mind, of the precariousness of the ground on which he was standing. On this ground, he never ventured a positive affirmation—but, resigning the lofty tone of demonstration, and putting on the modesty of conscious ignorance, he brought forward all he had to say in the humble form of a doubt, or a conjecture, or a question. But, what he had not confidence to do, other philosophers have done after him—and they have winged their audacious way into forbidden regions—and they have crossed that circle by which the field of observation is inclosed—and there have they debated and dogmatised with all the pride of a most intolerant assurance.

Now, though the case be imaginary, let us conceive, for the sake of illustration, that one of these philosophers made so extravagant a departure from the sobriety of experimental science, as to pass on from the astronomy of the different planets, and to attempt the natural history of their animal and vegetable kingdoms. He might get hold of some vague and general analogies, to throw an air of plausibility around his speculation. He might pass from the botany of the different regions of the globe that we, inhabit, and make his loose and confident applications to each of the other planets, according to its distance from the sun, and the inclination of its axis to the plane of its annual revolution; and out of some such slender materials, he may work up an amusing philosophical romance, full of ingenuity, and having, withal, the colour of truth and of consistency spread over it.

I can conceive how a superficial public might be delighted by the eloquence of such a composition, and even be impressed by its arguments; but were I asked, which is the man of all the ages and countries in the world, who would have the least respect for this treatise upon the plants which grow on the surface of Jupiter, I should be at no loss to answer the question. I should say, that it would be he who had computed the motions of Jupiter—that it would be he who had measured the bulk and the density of Jupiter—that it would be he who had estimated the periods of Jupiter—that it would be he whose observant eye and patiently calculating mind, had traced the satellites of Jupiter through all the rounds of their mazy circulation, and unravelled the intricacy of all their movements. He would see at once that the subject lay at a hopeless distance beyond the field of legitimate observation. It would be quite enough for him, that it was beyond the range of his telescope. On this ground, and on this ground only, would he reject it as one of the puniest imbecilities of childhood. As to any character of truth or of importance, it would have no more effect on such a mind as that of Newton, than any illusion of poetry; and from the eminence of his intellectual throne, would he cast a penetrating glance at the whole speculation, and bid its gaudy insignificance away from him.

But let us pass onward to another case, which though as imaginary as the former, may still serve the purpose of illustration.

This same adventurous philosopher may be conceived to shift his speculation from the plants of another world, to the character of its inhabitants. He may avail himself of some slender correspondencies between the heat of the sun and the moral temperament of the people it shines upon. He may work up a theory, which carries on the front of it some of the characters of plausibility; but surely it does not require the philosophy of Newton to demonstrate the folly of such an enterprise. There is not a man of plain understanding, who does not perceive that this said ambitious inquirer has got without his reach—that he has stepped beyond the field of experience, and is now expatiating on the field of imagination—that he has ventured on a dark unknown, where the wisest of all philosophy, is the philosophy of silence, and a profession of ignorance is the best evidence of a solid understanding—that if he think he knows any thing on such a subject as this, he knoweth nothing yet as he ought to know. He knows not what Newton knew, and what he kept a steady eye upon throughout the whole march of his sublime investigations. He knows not the limit of his own faculties. He has overleaped the barrier which hems in all the possibilities of human attainment. He has wantonly flung himself off from the safe and firm field of observation, and got on that undiscoverable ground, where, by every step he takes, he widens his distance from the true philosophy, and by every affirmation he utters, he rebels against the authority of all its maxims.

I can conceive it the feeling of every one of you, that I have hitherto indulged in a vain expense of argument, and it is most natural for you to put the question, "What is the precise point of convergence to which I am directing all the light of this abundant and seemingly superfluous illustration?"

In the astronomical objection which Infidelity has proposed against the truth of the Christian revelation, there is first an assertion, and then an argument. The assertion is, that Christianity is set up for the exclusive benefit of our minute and solitary world. The argument is, that God would not lavish such a quantity of attention on so insignificant a field. Even though the assertion were admitted, I should have a quarrel with the argument. But the futility of the objection is not laid open in all its extent, unless we expose the utter want of all essential evidence even for the truth of the assertion. How do infidels know that Christianity is set up for the single benefit of this earth and its inhabitants? How are they able to tell us, that if you go to other planets, the person and the religion of Jesus are there unknown to them? We challenge them to the proof of this said positive announcement of theirs. We see in this objection the same rash and gratuitous procedure, which was so apparent in the two cases that we have already advanced for the purpose of illustration. We see in it the same glaring transgression on the spirit and the maxims of that very philosophy which they profess to idolize. They have made their argument against us, out of an assertion which has positively no feet to rest upon—an assertion which they have no means whatever of verifying—an assertion, the truth or the falsehood of which can only be gathered out of some supernatural message, for it lies completely beyond the range of human observation. It is willingly admitted, that by an attempt at the botany of other worlds, the true method of philosophising is trampled on; for this is a subject that lies beyond the range of actual observation, and every performance upon it must be made up of assertions without proofs. It is also willingly admitted, that an attempt at the civil and political history of their people, would be an equally extravagant departure from the spirit of the true philosophy; for this also lies beyond the field of actual observation; and all that could possibly be mustered up on such a subject as this, would still be assertions without proofs. Now, the theology of these planets is, in every way, as inaccessible a subject as their politics or their natural history; and therefore it is, that the objection, grounded on the confident assumption of those infidel astronomers, who assert Christianity to be the religion of this one world, or that the religion of these other worlds is not our very Christianity, can have no influence on a mind that has derived its habits of thinking, from the pure and rigorous school of Newton; for the whole of this assertion is just as glaringly destitute, as in the two former instances, of proof.

The man who could embark in an enterprise so foolish and so fanciful, as to theorise it on the details of the botany of another world, or to theorise it on the natural and moral history of its people, is just making as outrageous a departure from all sense, and all science, and all sobriety, when he presumes to speculate, or to assert on the details or the methods of God's administration among its rational and accountable inhabitants. He wings his fancy to as hazardous a region, and vainly strives a penetrating vision through the mantle of as deep an obscurity. All the elements of such a speculation are hidden from him. For any thing he can tell, sin has found its way into these other worlds. For any thing he can tell, their people have banished themselves from communion with God. For any thing he can tell, many a visit has been made to each of them, on the subject of our common Christianity, by commissioned messengers from the throne of the Eternal. For any thing he can tell, the redemption proclaimed to us is not one solitary instance, or not the whole of that redemption which is by the Son of God—but only our part in a plan of mercy, equal in magnificence to all that astronomy has brought within the range of human contemplation. For any thing he can tell, the moral pestilence, which walks abroad over the face of our world, may have spread its desolation over all the planets of all the systems, which the telescope has made known to us. For any thing he can tell, some mighty redemption has been devised in heaven, to meet this disaster in the whole extent and malignity of its visitations. For any thing he can tell, the wonder-working God, who has strewed the field of immensity with so many worlds, and spread the shelter of his omnipotence over them, may have sent a message of love to each, and re-assured the hearts of its despairing people by some overpowering manifestation of tenderness. For any thing he can tell, angels from paradise may have sped to every planet their delegated way, and sung, from each azure canopy, a joyful annunciation, and said, "Peace be to this residence, and good will to all its families, and glory to Him in the highest, who, from the eminency of his throne, has issued an act of grace so magnificent, as to carry the tidings of life and of acceptance to the unnumbered orbs of a sinful creation." For any thing he can tell, the Eternal Son, of whom it is said, that by him the worlds were created, may have had the government of many sinful worlds laid upon his shoulders; and by the power of his mysterious word, have awoke them all from that spiritual death, to which they had sunk in lethargy as profound as the slumbers of non-existence. For any thing he can tell, the one Spirit who moved on the face of the waters, and whose presiding influence it was that hushed the wild war of nature's elements, and made a beauteous system emerge out of its disjointed materials, may now be working with the fragments of another chaos; and educing order, and obedience, and harmony, out of the wrecks of a moral rebellion, which reaches through all these spheres, and spreads disorder to the uttermost limits of our astronomy.

But, here I stop—nor shall I attempt to

grope my dark and fatiguing way, by another inch, among such sublime and mysterious secrecies. It is not I who am offering to lift this curtain. It is not I who am pitching my adventurous flight to the secret things, which belong to God, away from the things that are revealed, and which belong to me and to my children. It is the champion of that very Infidelity which I am now combating. It is he who props his unchristian argument, by presumptions fetched out of those untravelled obscurities, which lie on the other side of a barrier that I pronounce to be impassable. It is he who transgresses the limits which Newton forbore to enter; because, with a justness which reigns throughout all his inquiries, he saw the limit of his own understanding, nor would he venture himself beyond it. It is he who has borrowed from the philosophy of this wondrous man, a few dazzling conceptions, which have only served to bewilder him—while, an utter stranger to the spirit of this philosophy, he has carried a daring and an ignorant speculation far beyond the boundary of its prescribed and allowable enterprises. It is he who has mustered against the truths of the Gospel, resting, as it does, on evidence within the reach of his faculties, an objection, for the truth of which he has no evidence whatever. It is he who puts away from him a doctrine, for which he has the substantial and the familiar proof of human testimony; and substitutes in its place, a doctrine, for which he can get no other support than from a reverie of his own imagination. It is he who turns aside from all that safe and certain argument, that is supplied by the history of this world, of which he knows something; and who loses himself in the work of theorising about other worlds, of the moral and theological history of which he positively knows nothing. Upon him, and not upon us, lies the folly of launching his impetuous way beyond the province of observation—of letting his fancy afloat among the unknown of distant and mysterious regions—and, by an act of daring, as impious as it is unphilosophical, of trying to unwrap that shroud, which, till drawn aside by the hand of a messenger from heaven, will ever veil, from human eye, the purposes of the Eternal.

If you have gone along with me in the preceding observations, you will perceive how they are calculated to disarm of all its point, and of all its energy, that flippancy of Voltaire; when, in the examples he gives of the dotage of the human understanding, he tells us of Bacon having believed in witchcraft, and Sir. Isaac Newton having written a Commentary on the Book of Revelation. The former instance we shall not undertake to vindicate; but in the latter instance, we perceive what this brilliant and specious, but withal superficial apostle of Infidelity, either did not see, or refused to acknowledge. We see in this intellectual labour of our great philosopher, the working of the very same principles which carried him through the profoundest and the most successful of his investigations; and how he kept most sacredly and most consistently by those very maxims, the authority of which, he, even in the full vigour and manhood of his faculties, ever recognised. We see in the theology of Newton, the very spirit and principle which gave all its stability, and all its sureness, to the philosophy of Newton. We see the same tenacious adherence to every one doctrine, that had such valid proof to uphold it, as could be gathered from the field of human experience; and we see the same firm resistance of every one argument, that had nothing to recommend it, but such plausibilities as could easily be devised by the genius of man, when he expatiated abroad on those fields of creation which the eye never witnessed, and from which no messenger ever came to us with any credible information. Now, it was on the former of these two principles that Newton clung so determinedly to his Bible, as the record of an actual annunciation from God to the inhabitants of this world. When he turned his attention to this book, he came to it with a mind tutored to the philosophy of facts—and, when he looked at its credentials, he saw the stamp and the impress of this philosophy on every one of them. He saw the fact of Christ being a messenger from heaven, in the audible language by which it was conveyed from heaven's canopy to human ears. He saw the fact of his being an approved ambassador of God, in those miracles which carried their own resistless evidence along with them to human eyes. He saw the truth of this whole history brought home to his own conviction, by a sound and substantial vehicle of human testimony. He saw the reality of that supernatural light, which inspired the prophecies he himself illustrated, by such an agreement with the events of a various and distant futurity as could be taken cognizance of by human observation. He saw the wisdom of God pervading the whole substance of the written message, in such manifold adaptations to the circumstances of man, and to the whole secrecy of his thoughts, and his affections, and his spiritual wants, and his moral sensibilities, as even in the mind of an ordinary and unlettered peasant, can be attested by human consciousness. These formed the solid materials of the basis on which our experimental philosopher stood; and there was nothing in the whole compass of his own astronomy, to dazzle him away from it; and he was too well aware of the limit between what he knew, and what he did not know, to be seduced from the ground he had taken, by any of those brilliancies, which have since led so many of his humbler successors into the track of Infidelity. He had measured the distances of these planets. He had calculated their periods. He had estimated their figures, and their bulk, and their densities, and he had subordinated the whole intricacy of their movements to the simple and sublime agency of one commanding principle. But he had too much of the ballast of a substantial understanding about him, to be thrown afloat by all this success among the plausibilities of wanton and unauthorised speculation. He knew the boundary which hemmed him. He knew that he had not thrown one particle of light on the moral or religious history of these planetary regions. He had not ascertained what visits of communication they received from the God who upholds them. But he knew that the fact of a real visit made to this planet, had such evidence to rest upon, that it was not to be disposted by any aerial imagination. And when I look at the steady and unmoved Christianity of this wonderful man; so far from seeing any symptom of dotage and imbecility, or any forgetfulness of those principles on which the fabric of his philosophy is reared; do I see, that in sitting down to the work of a Bible Commentator, he hath given us their most beautiful and most consistent exemplification.

I did not anticipate such a length of time, and of illustration, in this stage of my argument. But I will not regret it, if I have familiarised the minds of any of my readers to the reigning principle of this Discourse. We are strongly disposed to think, that it is a principle which might be made to apply to every argument of every unbeliever—and so to serve not merely as an antidote against the infidelity of astronomers, but to serve as an antidote against all Infidelity. We are well aware of the diversity of complexion which Infidelity puts on. It looks one thing in the man of science and of liberal accomplishment. It looks another thing in the refined voluptuary. It looks still another thing in the common-place railer against the artifices of priestly domination. It looks another thing in the dark and unsettled spirit of him, whose every reflection is tinctured with gall, and who casts his envious and malignant scowl at all that stands associated with the established order of society. It looks another thing in the prosperous man of business, who has neither time nor patience for the details of the Christian evidence—but who, amid the hurry of his other occupations, has gathered as many of the lighter petulancies of the infidel writers, and caught, from the perusal of them, as contemptuous a tone towards the religion of the New Testament, as to set him at large from all the decencies of religious observation, and to give him the disdain of an elevated complacency over all the follies of what he counts a vulgar superstition. And, lastly, for Infidelity has now got down amongst us to the humblest walks of life; may it occasionally be seen louring on the forehead of the resolute and hardy artificer, who can lift his menacing voice against the priesthood, and, looking on the Bible as a jugglery of theirs, can bid stout defiance to all its denunciations. Now, under all these varieties, we think that there might be detected the one and universal principle which we have attempted to expose. The something, whatever it is, which has dispossessed all these people of their Christianity, exists in their minds, in the shape of a position, which they hold to be true, but which, by no legitimate evidence, they have ever realised—and a position, which lodges within them as a wilful fancy or presumption of their own, but which could not stand the touchstone of that wise and solid principle, in virtue of which, the followers of Newton give to observation the precedence over theory. It is a principle altogether worthy of being laboured—as, if carried round in faithful and consistent application amongst these numerous varieties, it is able to break up all the existing Infidelity of the world.

But, there is one other most important conclusion, to which it carries us. It carries us, with all the docility of children, to the Bible; and puts us down into the attitude of an unreserved surrender of thought and understanding, to its authoritative information. Without the testimony of an authentic messenger from heaven, I know nothing of heaven's counsels. I never heard of any moral telescope that can bring to my observation, the doings or the deliberations which are taking place in the sanctuary of the Eternal. I may put into the registers of my belief, all that comes home to me through the senses of the outer man, or by the consciousness of the inner man. But neither the one nor the other can tell me of the purposes of God; can tell me of the transactions or the designs of his sublime monarchy; can tell me of the goings forth of Him who is from everlasting unto everlasting; can tell me of the march and the movements of that great administration which embraces all worlds, and takes into its wide and comprehensive survey, the mighty roll of innumerable ages. It is true that my fancy may break its impetuous way into this lofty and inaccessible field; and through the devices of my heart, which are many, the visions of an ever-shifting theology may take their alternate sway over me; but the counsel of the Lord, it shall stand. And I repeat it, that if true to the leading principle of that philosophy, which has poured such a flood of light over the mysteries of nature, we shall dismiss every self-formed conception of our own, and wait in all the humility of conscious ignorance, till the Lord himself shall break his silence, and make his counsel known, by an act of communication.

And now, that a professed communication is before me, and that it has all the solidity of the experimental evidence on its side, and nothing but the reveries of a daring speculation to oppose it, what is the consistent, what is the rational; what is the philosophical use that should be made of this document, but to set me down like a school-boy, to the work of turning its pages, and conning its lessons, and submitting the every exercise of my judgment to its information and its testimony? We know that there is a superficial philosophy, which casts the glare of a most seducing brilliancy around it; and spurns the Bible, with all the doctrine, and all the piety of the Bible, away from it; and has infused the spirit of Antichrist into many of the literary establishments of the age; but it is not the solid, the profound, the cautious spirit of that philosophy, which has done so much to ennoble the modern period of our world; for the more that this spirit is cultivated and understood, the more will it be found in alliance with that spirit, in virtue of which all that exalteth itself against the knowledge of God, is humbled, and all lofty imaginations are cast down, and every thought of the heart is brought into the captivity of the obedience of Christ.

DISCOURSE III.

ON THE EXTENT OF THE DIVINE CONDESCENSION.

"Who is like unto the Lord our God, who dwelleth on high. Who humbleth himself to behold the things that are in heaven, and in the earth!"—Psalm cxiii. 5, 6.

In our last Discourse, we attempted to expose the total want of evidence for the assertion of the infidel astronomer—and this reduces the whole of our remaining controversy with him, to the business of arguing against a mere possibility. Still, however, the answer is not so complete as it might be, till the soundness of the argument be attended to, as well as the credibility of the assertion—or, in other words, let us admit the assertion, and take a view of the reasoning which has been constructed upon it.

We have already attempted to lay before you, the wonderful extent of that space, teeming with unnumbered worlds, which modern science has brought within the circle of its discoveries. We even ventured to expatiate on those tracks of infinity, which lie on the other side of all that eye or that telescope hath made known to us—to shoot afar into those ulterior regions, which are beyond the limits of our astronomy—to impress you with the rashness of the imagination, that the creative energy of God had sunk exhausted by the magnitude of its efforts, at that very line, through which the art of man, lavished as it has been on the work of perfecting the instruments of vision, has not yet been able to penetrate; and upon all this we hazarded the assertion, that though all these visible heavens were to rush into annihilation, and the besom of the Almighty's wrath were to sweep from the face of the universe, those millions, and millions more of suns and of systems, which lie within the grasp of our actual observation—that this event, which, to our eye, would leave so wide and so dismal a solitude behind it, might be nothing in the eye of Him who could take in the whole, but the disappearance of a little speck from that field of created things, which the hand of his omnipotence had thrown around him.

But to press home the sentiment of the text, it is not necessary to stretch the imagination beyond the limit of our actual discoveries. It is enough to strike our minds with the insignificance of this world, and of all who inhabit it, to bring it into measurement with that mighty assemblage of worlds, which lie open to the eye of man, aided as it has been by the inventions of his genius. When we told you of the eighty millions of suns, each occupying his own independent territory in space, and dispensing his own influences over a cluster of tributary worlds; this world could not fail to sink into littleness in the eye of him, who looked to all the magnitude and variety which are around it. We gave you but a feeble image of our comparative insignificance, when we said that the glories of an extended forest, would suffer no more from the fall of a single leaf, than the glories of this extended universe would suffer, though the globe we tread, "and all that it inherit, should dissolve." And when we lift our conceptions to Him who has peopled immensity with all these wonders—who sits enthroned on the magnificence of his own works, and by one sublime idea can embrace the whole extent of that boundless amplitude, which he has filled with the trophies of his divinity; we cannot but resign our whole heart to the Psalmist's exclamation of "What is man, that thou art mindful of him, or the son of man, that thou shouldest deign to visit him!"

Now mark the use to which all this has been turned by the genius of Infidelity. Such a humble portion of the universe as ours, could never have been the object of such high and distinguishing attentions as Christianity has assigned to it. God would not have manifested himself in the flesh for the salvation of so paltry a world. The monarch of a whole continent, would never move from his capital; and lay aside the splendour of royalty; and subject himself for months, or for years, to perils, and poverty, and persecution; and take up his abode in some small islet of his dominions, which, though swallowed by an earthquake, could not be missed amid the glories of so wide an empire; and all this to regain the lost affections of a few families upon its surface. And neither would the eternal Son of God—he who is revealed to us as having made all worlds, and as holding an empire, amid the splendours of which the globe that we inherit, is shaded in insignificance; neither would he strip himself of the glory he had with the Father before the world was, and light on this lower scene, for the purpose imputed to him in the New Testament. Impossible, that the concerns of this puny ball, which floats its little round among an infinity of larger worlds, should be of such mighty account in the plans of the Eternal, or should have given birth in heaven to so wonderful a movement, as the Son of God putting on the form of our degraded species, and sojourning amongst us, and sharing in all our infirmities, and crowning the whole scene of humiliation, by the disgrace and the agonies of a cruel martyrdom.

This has been started as a difficulty in the way of the Christian Revelation; and it is the boast of many of our philosophical Infidels, that by the light of modern discovery, the light of the New Testament is eclipsed and overborne; and the mischief is not confined to philosophers, for the argument has got into other hands, and the popular illustrations that are now given to the sublimest truths of science, have widely disseminated all the Deism that has been grafted upon it; and the high tone of a decided contempt for the Gospel, is now associated with the flippancy of superficial acquirements; and, while the venerable Newton, whose genius threw open those mighty fields of contemplation, found a fit exercise for his powers in the interpretation of the Bible, there are thousands and tens of thousands, who, though walking in the light which he holds out to them, are seduced by a complacency which he never felt, and inflated by a pride which never entered into his pious and philosophical bosom, and whose only notice of the Bible, is to depreciate, and to deride, and to disown it.

Before entering into what we conceive to be the right answer to this objection, let us previously observe, that it goes to strip the Deity of an attribute, which forms a wonderful addition to the glories of his incomprehensible character. It is indeed a mighty evidence of the strength of his arm, that so many millions of worlds are suspended on it; but it would surely make the high attribute of his power more illustrious, if, while it expatiated at large among the suns and the systems of astronomy, it could, at the very same instant, be impressing a movement and a direction on all the minuter wheels of that machinery, which is working incessantly around us. It forms a noble demonstration of his wisdom, that he gives unremitting operation to those laws which uphold the stability of this great universe; but it would go to heighten that wisdom inconceivably, if, while equal to the magnificent task of maintaining the order and harmony of the spheres, it was lavishing its inexhaustible resources on the beauties, and varieties, and arrangements, of every one scene, however humble, of every one field, however narrow, of the creation he had formed. It is a cheering evidence of the delight he takes in communicating happiness, that the whole of immensity should be so strewed with the habitations of life and of intelligence; but it would surely bring home the evidence, with a nearer and a more affecting impression, to every bosom, did we know, that at the very time his benignant regard took in the mighty circle of created beings, there was not a single family overlooked by him, and that every individual in every corner of his dominions, was as effectually seen to, as if the object of an exclusive and undivided care. It is our imperfection, that we cannot give our attention to more than one object, at one and the same instant of time; but surely it would elevate our every idea of the perfections of God, did we know, that while his comprehensive mind could grasp the whole amplitude of nature, to the very outermost of its boundaries, he had an attentive eye fastened on the very humblest of its objects, and pondered every thought of my heart, and noticed every footstep of my goings, and treasured up in his remembrance every turn and every movement of my history.

And, lastly, to apply this train of sentiment to the matter before us; let us suppose that one among the countless myriads of worlds, should be visited by a moral pestilence, which spread through all its people, and brought them under the doom of a law, whose sanctions were unrelenting and immutable; it were no disparagement to God, should he, by an act of righteous indignation, sweep this offence away from the universe which it deformed—nor should we wonder, though, among the multitude of other worlds, from which the ear of the Almighty was regaled with the songs of praise, and the incense of a pure adoration ascended to his throne, he should leave the strayed and solitary world to perish in the guilt of its rebellion. But, tell me, oh! tell me, would it not throw the softening of a most exquisite tenderness over the character of God, should we see him putting forth his every expedient to reclaim to himself those children who had wandered away from him—and, few as they were when compared with the host of his obedient worshippers, would it not just impart to his attribute of compassion the infinity of the Godhead, that, rather than lose the single world which had turned to its own way, he should send the messengers of peace to woo and to welcome it back again; and, if justice demanded so mighty a sacrifice, and the law behoved to be so magnified and made honourable, tell me whether it would not throw a moral sublime over the goodness of the Deity, should he lay upon his own Son the burden of its atonement, that he might again smile upon the world, and hold out the sceptre of invitation to all its families?

We avow it, therefore, that this infidel argument goes to expunge a perfection from the character of God. The more we know of the extent of nature, should not we have the loftier conception of him who sits in high authority over the concerns of so wide a universe? But, is it not adding to the bright catalogue of his other attributes, to say, that, while magnitude does not overpower him, minuteness cannot escape him, and variety cannot bewilder him; and that, at the very time while the mind of the Deity is abroad over the whole vastness of creation, there is not one particle of matter, there is not one individual principle of rational or of animal existence, there is not one single world in that expanse which teems with them, that his eye does not discern as constantly, and his hand does not guide as unerringly, and his spirit does not watch and care for as vigilantly, as if it formed the one and exclusive object of his attention.

The thing is inconceivable to us, whose minds are so easily distracted by a number of objects, and this is the secret principle of the whole Infidelity I am now alluding to. To bring God to the level of our own comprehension, we would clothe him in the impotency of a man. We would transfer to his wonderful mind all the imperfection of our own faculties. When we are taught by astronomy, that he has millions of worlds to look after, and thus add in one direction to the glories of his character; we take away from them in another, by saying, that each of these worlds must be looked after imperfectly. The use that we make of a discovery, which should heighten our every conception of God, and humble us into the sentiment, that a Being of such mysterious elevation is to us unfathomable, is to sit in judgment over him, aye, and to pronounce such a judgment as degrades him, and keeps him down to the standard of our own paltry imagination! We are introduced by modern science to a multitude of other suns and of other systems; and the perverse interpretation we put upon the fact, that God *can* diffuse the benefits of his power and of his goodness over such a variety of worlds, is, that he *cannot*, or will not, bestow so much goodness on one of those worlds, as a professed revelation from Heaven has announced to us. While we enlarge the provinces of his empire, we tarnish all the glory of this enlargement, by saying, he has so much to care for, that the care of every one province must be less complete, and less vigilant, and less effectual, than it would otherwise have been. By the discoveries of modern science, we multiply the places of the creation; but along with this, we would impair the attribute of his eye being in every place to behold the evil and the good; and thus, while we magnify one of his perfections, we do it at the expense of another; and to bring him within the grasp of our feeble capacity, we would deface one of the glories of that character, which it is our part to adore, as higher than all thought, and as greater than all comprehension.

The objection we are discussing, I shall state again in a single sentence. Since astronomy has unfolded to us such a number of worlds, it is not likely that God would pay so much attention to this one world, and set up such wonderful provisions for its benefit, as are announced to us in the Christian Revelation. This objection will have received its answer, if we can meet it by the following position:—that God, in addition to the bare faculty of dwelling on a multiplicity of objects at one and the same time, has this faculty in such wonderful perfection, that he can attend as fully, and provide as richly, and manifest all his attributes as illustriously, on every one of these objects, as if the rest had no existence, and no place whatever in his government or in his thoughts.

For the evidence of this position, we appeal, in the first place, to the personal history of each individual among you. Only grant us, that God never loses sight of any one thing he has created, and that no created thing can continue either to be, or to act independently of him; and then, even upon the face of this world, humble as it is on the great scale of astronomy, how widely diversified, and how multiplied into many thousand distinct exercises, is the attention of God! His eye is upon every hour of my existence. His Spirit is intimately present with every thought of my heart. His inspiration gives birth to every purpose within me. His hand impresses a direction on every footstep of my goings. Every breath I inhale, is drawn by an energy which God deals out to me. This body, which, upon the slightest derangement, would become the prey of death, or of woeful suffering, is now at ease, because he at this moment is warding off from me a thousand dangers, and upholding the thousand movements of its complex and delicate machinery. His presiding influence keeps by me through the whole current of my restless and ever-changing history. When I walk by the way-side, he is along with me. When I enter into company, amid all my forgetfulness of him, he never forgets me. In the silent watches of the night, when my eyelids have closed, and my spirit has sunk into unconsciousness, the observant eye of him who never slumbers, is upon me. I cannot fly from his presence. Go where I will, he tends me, and watches me, and cares for me; and the same Being who is now at work in the remotest domains of Nature and of Providence, is also at my right hand to eke out to me every moment of my being, and to uphold me in the exercise of all my feelings, and of all my faculties.

Now, what God is doing with me, he is doing with every distinct individual of this world's population. The intimacy of his presence, and attention, and care, reaches to one and to all of them. With a mind unburdened by the vastness of all its other concerns, he can prosecute, without distraction, the government and guardianship of every one son and daughter of the species.—And is it for us, in the face of all this experience, ungratefully to draw a limit around the perfections of God—to aver, that the multitude of other worlds has withdrawn any portion of his benevolence from the one we occupy—or that he, whose eye is upon every separate family of the earth, would not lavish all the riches of his unsearchahle attributes on some high plan of pardon and immortality, in behalf of its countless generations?

But, secondly, were the mind of God so fatigued, and so occupied with the care of other worlds, as the objection presumes him to be, should we not see some traces of neglect, or of carelessness, in his management of ours? Should we not behold, in many a field of observation, the evidence of its master being overcrowded with the variety of his other engagements? A man oppressed by a multitude of business, would simplify and reduce the work of any new concern that was devolved upon him. Now, point out a single mark of God being thus oppressed. Astronomy has laid open to us so many realms of creation, which were before unheard of, that the world we inhabit, shrinks into one remote and solitary province of his wide monarchy. Tell me, then, if, in any one field of this province, which man has access to, you witness a single indication of God sparing himself—of God reduced to languor by the weight of his other employments—of God sinking under the burden of that vast superintendence which lies upon him—of God being exhausted, as one of ourselves would be, by any number of concerns, however great, by any variety of them, however manifold; and do you not perceive, in that mighty profusion of wisdom and of goodness, which is scattered every where around us, that the thoughts of this unsearchable Being are not as our thoughts, nor his ways as our ways?

My time does not suffer me to dwell on this topic, because, before I conclude, I must hasten to another illustration. But, when I look abroad on the wondrous scene that is immediately before me—and see, that in every direction, it is a scene of the most various and unwearied activity—and expatiate on all the beauties of that garniture by which it is adorned, and on all the prints of design and of benevolence which abound in it—and think, that the same

God, who holds the universe, with its every system, in the hollow of his hand, pencils every flower, and gives nourishment to every blade of grass, and actuates the movements of every living thing, and is not disabled, by the weight of his other cares, from enriching the humble department of nature I occupy, with charms and accommodations of the most unbounded variety—then, surely, if a message, bearing every mark of authenticity, should profess to come to me from God, and inform me of his mighty doings for the happiness of our species, it is not for me, in the face of all this evidence, to reject it as a tale of imposture, because astronomers have told me that he has so many other worlds and other orders of beings to attend to—and, when I think that it were a deposition of him from his supremacy over the creatures he has formed, should a single sparrow fall to the ground without his appointment, then let science and sophistry try to cheat me of my comfort as they may—I will not let go the anchor of my confidence in God—I will not be afraid, for I am of more value than many sparrows.

But, thirdly, it was the telescope, that, by piercing the obscurity which lies between us and distant worlds, put Infidelity in possession of the argument, against which we are now contending. But, about the time of its invention, another instrument was formed, which laid open a scene no less wonderful, and rewarded the inquisitive spirit of man with a discovery, which serves to neutralise the whole of this argument. This was the microscope. The one led me to see a system in every star. The other leads me to see a world in every atom. The one taught me, that this mighty globe, with the whole burden of its people, and of its countries, is but a grain of sand on the high field of immensity. The other teaches me, that every grain of sand may harbour within it the tribes and the families of a busy population. The one told me of the insignificance of the world I tread upon. The other redeems it from all its insignificance; for it tells me that in the leaves of every forest, and in the flowers of every garden, and in the waters of every rivulet, there are worlds teeming with life, and numberless as are the glories of the firmament.

The one has suggested to me, that beyond and above all that is visible to man, there may lie fields of creation which sweep immeasurably along, and carry the impress of the Almighty's hand to the remotest scenes of the universe. The other suggests to me, that within and beneath all that minuteness which the aided eye of man has been able to explore, there may be a region of invisibles; and that could we draw aside the mysterious curtain which shrouds it from our senses, we might there see a theatre of as many wonders as astronomy has unfolded, a universe within the compass of a point so small, as to elude all the powers of the microscope, but where the wonder-working God finds room for the exercise of all his attributes, where he can raise another mechanism of worlds, and fill and animate them all with the evidences of his glory.

Now, mark how all this may be made to meet the argument of our infidel astronomer. By the telescope they have discovered, that no magnitude, however vast, is beyond the grasp of the Divinity. But by the microscope, we have also discovered, that no minuteness, however shrunk from the notice of the human eye, is beneath the condescension of his regard. Every addition to the powers of the one instrument, extends the limit of his visible dominions. But, by every addition to the powers of the other instrument, we see each part of them more crowded than before, with the wonders of his unwearying hand. The one is constantly widening the circle of his territory. The other is as constantly filling up its separate portions, with all that is rich, and various, and exquisite. In a word, by the one I am told that the Almighty is now at work in regions more distant than geometry has ever measured, and among worlds more manifold than numbers have ever reached. But, by the other, I am also told, that, with a mind to comprehend the whole, in the vast compass of its generality, he has also a mind to concentrate a close and a separate attention on each and on all of its particulars; and that the same God, who sends forth an upholding influence among the orbs and the movements of astronomy, can fill the recesses of every single atom with the intimacy of his presence, and travel, in all the greatness of his unimpaired attributes, upon every one spot and corner of the universe he has formed.

They, therefore, who think that God will not put forth such a power, and such a goodness, and such a condescension, in behalf of this world, as are ascribed to him in the New Testament, because he has so many other worlds to attend to, think of him as a man. They confine their view to the informations of the telescope, and forget altogether the informations of the other instrument. They only find room in their minds for his one attribute of a large and general superintendence, and keep out of their remembrance the equally impressive proofs we have for his other attribute of a minute and multiplied attention to all that diversity of operations, where it is he that worketh all in all. And when I think, that, as one of the instruments of philosophy has heightened our every impression of the first of these attributes, so another instrument has no less heightened our impression of the second of them—then I can no longer resist the conclusion, that it would be a transgression of sound argument, as well as a daring of impiety, to draw a limit around the doings of this unsearchable God— and, should a professed revelation from heaven, tell me of an act of condescension, in behalf of some separate world, so wonderful, that angels desired to look into it, and the Eternal Son had to move from his seat of glory to carry it into accomplishment, all I ask is the evidence of such a revelation; for, let it tell me as much as it may of God letting himself down for the benefit of one single province of his dominions, this is no more than what I see lying scattered, in numberless examples, before me; and running through the whole line of my recollections; and meeting me in every walk of observation to which I can betake myself; and, now that the microscope has unveiled the wonders of another region, I see strewed around me, with a profusion which baffles my every attempt to comprehend it, the evidence that there is no one portion of the universe of God too minute for his notice, nor too humble for the visitations of his care.

As the end of all these illustrations, let me bestow a single paragraph on what I conceive to be the precise state of this argument.

It is a wonderful thing that God should be so unencumbered by the concerns of a whole universe, that he can give a constant attention to every moment of every individual in this world's population. But, wonderful as it is, you do not hesitate to admit it as true, on the evidence of your own recollections. It is a wonderful thing that he whose eye is at every instant on so many worlds, should have peopled the world we inhabit with all the traces of the varied design and benevolence which abound in it. But, great as the wonder is, you do not allow so much as the shadow of improbability to darken it, for its reality is what you actually witness, and you never think of questioning the evidence of observation. It is wonderful, it is passing wonderful, that the same God, whose presence is diffused through immensity, and who spreads the ample canopy of his administration over all its dwelling-places, should, with an energy as fresh and as unexpended as if he had only begun the work of creation, turn him to the neighbourhood around us, and lavish, on its every hand-breadth, all the exuberance of his goodness, and crowd it with the many thousand varieties of conscious existence. But, be the wonder incomprehensible as it may, you do not suffer in your mind the burden of a single doubt to lie upon it, because you do not question the report of the microscope. You do not refuse its information, nor turn away from it as an incompetent channel of evidence. But to bring it still nearer to the point at issue, there are many who never looked through a microscope, but who rest an implicit faith in all its revelations; and upon what evidence I would ask? Upon the evidence of testimony—upon the credit they give to the authors of the books they have read, and the belief they put in the record of their observations. Now, at this point I make my stand. It is wonderful that God should be so interested in the redemption of a single world, as to send forth his well-beloved Son upon the errand, and he, to accomplish it, should, mighty to save, put forth all his strength, and travail in the greatness of it. But such wonders as these have already multiplied upon you; and when evidence is given of their truth, you have resigned your every judgment of the unsearchable God, and rested in the faith of them. I demand, in the name of sound and consistent philosophy, that you do the same in the matter before us—and take it up as a question of evidence—and examine that medium of testimony through which the miracles and informations of the Gospel have come to your door—and go not to admit as argument here, what would not be admitted as argument in any of the analogies of nature and observation—and take along with you in this field of inquiry, a lesson which you should have learned upon other fields—even the depth of the riches both of the wisdom and the knowledge of God, that his judgments are unsearchable, and his ways are past finding out.

I do not enter at all into the positive evidence for the truth of the Christian Revelation, my single aim at present being to dispose of one of the objections which is conceived to stand in the way of it. Let me suppose then that this is done to the satisfaction of a philosophical inquirer, and that the evidence is sustained, and that the same mind that is familiarised to all the sublimities of natural science, and has been in the habit of contemplating God in association with all the magnificence which is around him, shall be brought to submit its thoughts to the captivity of the doctrine of Christ. Oh! with what veneration, and gratitude, and wonder, should he look on the descent of him into this lower world, who made all these things, and without whom was not any thing made that was made. What a grandeur does it throw over every step in the redemption of a fallen world, to think of its being done by him who unrobed him of the glories of so wide a monarchy, and came to this humblest of its provinces, in the disguise of a servant, and took upon him the form of our degraded species, and let himself down to sorrows, and to sufferings, and to death, for us. In this love of an expiring Saviour to those for whom in agony he poured out his soul, there is a height, and a depth, and a length, and a breadth, more than I can comprehend; and let me never never from this moment neglect so great a salvation, or lose my hold of an atonement, made sure by him who cried, that it was finished, and brought in an everlasting righteousness. It was not the visit of an empty parade that he made to us. It was for the accomplishment of some substantial purpose; and, if that purpose is announced, and stated to consist in his dying the just for the unjust, that he might bring us unto God, let us never doubt of our acceptance in that way of communication with our Father in heaven, which he hath opened and made known to us. In taking to that way, let us follow his every direction with that humility which a sense of all this wonderful condescension is fitted to inspire. Let us forsake all that he bids us forsake. Let us do all that he bids us do. Let us give ourselves up to his guidance with the docility of children, overpowered by a kindness that we never merited, and a love that is unequalled by all the perverseness and all the ingratitude of our stubborn nature—for what shall we render unto him for such mysterious benefits—to him who has thus been mindful of us—to him who thus has deigned to visit us?

But the whole of this argument is not yet exhausted. We have scarcely entered on the defence that is commonly made against the plea which Infidelity rests on the wonderful extent of the universe of God, and the insignificancy of our assigned portion of it. The way in which we have attempted to dispose of this plea, is by insisting on the evidence that is every where around us, of God combining with the largeness of a vast and mighty superintendence, which reaches the outskirts of creation, and spreads over all its amplitudes—the faculty of bestowing as much attention, and exercising as complete and manifold a wisdom, and lavishing as profuse and inexhaustible a goodness, on each of its humblest departments, as if it formed the whole extent of his territory.

In the whole of this argument we have looked upon the earth as isolated from the rest of the universe altogether. But according to the way in which the astronomical objection is commonly met, the earth is not viewed as in a state of detachment from the other worlds, and the other orders of being which God has called into existence. It is looked upon as the member of a more extended system. It is associated with the magnificence of a moral empire, as wide as the kingdom of nature. It is not merely asserted, what in our last Discourse has been already done, that for any thing we can know by reason, the plan of redemption may have its influences and its bearings on those creatures of God who people other regions, and occupy other fields in the immensity of his dominions; that to argue, therefore, on this plan being instituted for the single benefit of the world we live in, and of the species to which we belong, is a mere presumption of the Infidel himself; and that the objection he rears on it, must fall to the ground, when the vanity of the presumption is exposed. The Christian apologist thinks he can go farther than this—that he can not merely expose the utter baselessness of the Infidel assertion, but that he has positive ground for erecting an opposite and a confronting assertion in its place—and that after having neutralised their position, by showing the entire absence of all observation in its behalf, he can pass on to the distinct and affirmative testimony of the Bible.

We do think that this lays open a very interesting track, not of wild and fanciful, but of most legitimate and sober-minded speculation. And anxious as we are to put every thing that bears upon the Christian argument, into all its lights; and fearless as we feel for the result of a most thorough sifting of it; and thinking as we do think it, the foulest scorn that any pigmy philosopher of the day should mince his ambiguous scepticism to a set of giddy and ignorant admirers, or that a half-learned and superficial public should associate with the Christian priesthood, the blindness and the bigotry of a sinking cause—with these feelings, we are not disposed to blink a single question that may be started on the subject of the Christian evidences. There is not one of its parts or bearings which needs the shelter of a disguise thrown over it. Let the priests of another faith ply their prudential expedients, and look so wise and so wary in the execution of them. But Christianity stands in a higher and a firmer attitude. The defensive armour of a shrinking or timid policy does not suit her. Hers is the naked majesty of truth; and with all the grandeur of age, but with none of its infirmities, has she come down to us, and gathered new strength from the battles she has won in the many controversies of many generations. With such a religion as this there is nothing to hide. All should be above boards. And the broadest light of day should be made fully and freely to circulate throughout all her secrecies. But secrets she has none. To her belong the frankness and the simplicity of conscious greatness; and whether she grapple it with the pride of philosophy, or stand in fronted opposition to the prejudices of the multitude, she does it upon her own strength, and spurns all the props and all the auxiliaries of superstition away from her.

DISCOURSE IV.

ON THE KNOWLEDGE OF MAN'S MORAL HISTORY IN THE DISTANT PLACES OF CREATION.

"Which things the angels desire to look into."

1 Peter i. 12.

There is a limit, across which man cannot carry any one of his perceptions, and from the ulterior of which he cannot gather a single observation to guide or to inform him. While he keeps by the objects which are near, he can get the knowledge of them conveyed to his mind through the ministry of several of the senses. He can feel a substance that is within reach of his hand. He can smell a flower that is presented to him. He can taste the food that is before him. He can hear a sound of certain pitch and intensity; and, so much does this sense of hearing widen his intercourse with external nature, that, from the distance of miles, it can bring him in an occasional intimation.

But of all the tracks of conveyance which God has been pleased to open up between the mind of man, and the theatre by which he is surrounded, there is none by which he so multiplies his acquaintance with the rich and the varied creation on every side of him, than by the organ of the eye. It is this which gives to man his loftiest command over the scenery of nature. It is this by which so broad a range of observation is submitted to him. It is this which enables him by the act of a single moment, to send an exploring look over the surface of an ample territory, to crowd his mind with the whole assembly of its objects, and to fill his vision with those countless hues which diversify and adorn it. It is this which carries him abroad over all that is sublime in the immensity of distance; which sets him as it were on an elevated platform, from whence he may cast a surveying glance over the arena of innumerable worlds; which spreads before him so mighty a province of contemplation, that the earth he inhabits, only appears to furnish him with the pedestal on which he may stand, and from which he may descry the wonders of all that magnificence, which the Divinity has poured so abundantly around him. It is by the narrow outlet of the eye, that the mind of man takes its excursive flight over those golden tracks, where, in all the exhaustlessness of creative wealth, lie scattered the suns and the systems of astronomy. But oh! how good a thing it is, and how becoming well, for the philosopher to be humble even amid the proudest march of human discovery, and the sublimest triumphs of the human understanding, when he thinks of that unscaled barrier, beyond which no power either of eye or of telescope, shall ever carry him; when he thinks that on the other side of it, there is a height, and a depth, and a length, and a breadth, to which the whole of this concave and visible firmament, dwindles into the insignificancy of an atom—and above all, how ready should he be to cast his every lofty imagination away from him, when he thinks of the God, who, on the simple foundation of his word, has reared the whole of this stately architecture, and by the force of his preserving hand, continues to uphold it; aye, and should the word again come out from him, that this earth shall pass away, and a portion of the heavens which are around it, shall again fall back into the annihilation from which he at first summoned them, what an impressive rebuke does it bring on the swelling vanity of science, to think that the whole field of its most ambitious enterprises may be swept away altogether, and there remain before the eye of him who sitteth on the throne, an untravelled immensity, which he hath filled with innumerable splendours, and over the whole face of which he hath inscribed the evidence of his high attributes, in all their might, and in all their manifestation.

But man has a great deal more to keep him humble of his understanding, than a mere sense of that boundary which skirts and which terminates the material field of his contemplations. He ought also to feel, how within that boundary, the vast majority of things is mysterious and unknown to him—that even in the inner chamber of his own consciousness, where so much lies hidden from the observation of others, there is also to himself a little world of incomprehensibles; that if stepping beyond the limits of this familiar home, he look no farther than to the members of his family, there is much in the cast and the colour of every mind that is above his powers of divination; that in proportion as he recedes from the centre of his own personal experience, there is a cloud of ignorance and secrecy, which spreads, and thickens, and throws a deep and impenetrable veil over the intricacies of every one department of human contemplation; that of all around him, his knowledge is naked and superficial, and confined to a few of those more conspicuous lineaments which strike upon his senses; that the whole face, both of nature and of society, presents him with questions which he cannot unriddle, and tells him how beneath the surface of all that the eye can rest upon, there lies the profoundness of a most unsearchable latency; aye, and should he in some lofty enterprise of thought, leave this world, and shoot afar into those tracks of speculation which astronomy has opened, should he, baffled by the mysteries which beset his every footstep upon earth, attempt an ambitious flight toward the mysteries of heaven—let him go, but let the justness of a pious and philosophical modesty go along with him—let him forget not, that from the moment his mind has taken its ascending way for a few little miles above the world he treads upon, his every sense abandons him but one—that number, and motion, and magnitude, and figure, make up all the bareness of its elementary informations—that these orbs have sent him scarce another message than told by their feeble glimmering upon his eye, the simple fact of their existence—that he sees not the landscape of other worlds—that he knows not the moral system of any one of them—nor athwart the long and trackless vacancy which lies between, does there fall upon his listening ear, the hum of their mighty populations.

But the knowledge which he cannot fetch up himself from the obscurity of this wondrous but untravelled scene, by the exercise of any one of his own senses, might be fetched to him by the testimony of a competent messenger. Conceive a native of one of these planetary mansions to light upon our world, and all we should require, would be, to be satisfied of his credentials, that we may tack our faith to every point of information he had to offer us. With the solitary exception of what we have been enabled to gather by the instruments of astronomy, there is not one of his communications about the place he came from, on which we possess any means at all of confronting him; and, therefore, could he only appear before us invested with the characters of truth, we should never think of any thing else than taking up the whole matter of his testimony just as he brought it to us.

It were well had a sound philosophy schooled its professing disciples to the same kind of acquiescence in another message, which has actually come to the world; and has told us of matters still more remote from every power of unaided observation; and has been sent from a more sublime and mysterious distance, even from that God of whom it is said, that " clouds and darkness are the habitation of his throne;" and treating of a theme so lofty and so inaccessible, as the counsels of that Eternal Spirit, "whose goings forth are of old, even from everlasting," challenges of man that he should submit his every thought to the authority of this high communication. Oh! had the philosophers of the day known as well as their great Master, how to draw the vigorous land-mark which verges the field of legitimate discovery, they should have seen when it is that philosophy becomes vain, and science is falsely so called; and how it is, that when philosophy is true to her principles, she shuts up her faithful votary to the Bible, and makes him willing to count all but loss, for the knowledge of Jesus Christ, and of him crucified.

But let it be well observed, that the object of this message is not to convey information to us about the state of these planetary regions. This is not the matter with which it is fraught. It is a message from the throne of God to this rebellious province of his dominions; and the purpose of it is, to reveal the fearful extent of our guilt and of our danger, and to lay before us the overtures of reconciliation. Were a similar message sent from the metropolis of a mighty empire, to one of its remote and revolutionary districts, we should not look to it for much information about the state or economy of the intermediate provinces. This were a departure from the topic on hand—though still there may chance to be some incidental allusions to the extent and resources of the whole monarchy, to the existence of a similar spirit of rebellion in other quarters of the land, or to the general principle of loyalty by which it was pervaded. Some casual references of this kind may be inserted in such a proclamation, or they may not—and it is with this precise feeling of ambiguity that we open the record of that embassy which has been sent us from heaven, to see if we can gather any thing there, about other places of the creation, to meet the objections of the infidel astronomer. But, while we pursue this object, let us have a care not to push the speculation beyond the limits of the written testimony; let us keep a just and a steady eye on the actual boundary of our knowledge, that, throughout every distinct step of our argument, we might preserve that chaste and unambitious spirit, which characterises the philosophy of him who explored these distant heavens, and, by the force of his genius, unravelled the secret of that wondrous mechanism which upholds them.

The informations of the Bible upon this subject, are of two sorts—that from which we confidently gather the fact, that the history of the redemption of our species is known in other and distant places of the creation—and that, from which we indistinctly guess at the fact, that the redemption itself may stretch beyond the limits of the world we occupy.

And, here it may shortly be adverted to, that, though we know little or nothing of the moral and theological economy of the other planets, we are not to infer, that the beings who occupy these widely extended regions, even though not higher than we in the scale of understanding, know little of ours. Our first parents, ere they committed that act by which they brought themselves and their posterity into the need of redemption, had frequent and familiar intercourse with God. He walked with them in the garden of paradise; and there did angels hold their habitual converse; and, should the same unblotted innocence which charmed and attracted these superior beings to the haunts of Eden, be perpetuated in every planet but our own, then might each of them be the scene of high and heavenly communications, and an open way for the messengers of God be kept up with them all, and their inhabitants be admitted to a share in the themes and contemplations of angels, and have their spirits exercised on those things, of which we are told that the angels desired to look into them; and thus, as we talk of the public mind of a city, or the public mind of an empire—by the well-frequented avenues of a free and ready circulation, a public mind might be formed throughout the whole extent of God's sinless and intelligent creation—and, just as we often read of the eyes of all Europe being turned to the one spot where some affair of eventful importance is going on, there might be the eyes of a whole universe turned to the one world, where rebellion against the Majesty of heaven had planted its standard; and for the re-admission of which within the circle of his fellowship, God, whose justice was inflexible, but whose mercy he had, by some plan of mysterious wisdom, made to rejoice over it, was putting forth all the might, and travailing in all the greatness of the attributes which belonged to him.

But, for the full understanding of this argument, it must be remarked, that, while in our exiled habitation, where all is darkness, and rebellion, and enmity, the creature engrosses every heart; and our affections, when they shift at all, only wander from one fleeting vanity to another, it is not so in the habitations of the unfallen. There, every desire and every movement is subordinated to God. He is seen in all that is formed, and in all that is spread around them—and, amid the fulness of that delight with which they expatiate over the good and the fair of this wondrous universe, the animating charm which pervades their every contemplation, is, that they behold, on each visible thing, the impress of the mind that conceived, and of the hand that made and that upholds it. Here, God is banished from the thoughts of every natural man, and, by a firm and constantly maintained act of usurpation, do the things of sense and of time wield an entire ascendancy. There, God is all in all. They walk in his light. They rejoice in the beatitudes of his presence. The veil is from off their eyes, and they see the character of a presiding Divinity in every scene, and in every event to which the Divinity has given birth. It is this which stamps a glory and an importance on the whole field of their contemplations; and when they see a new evolution in the history of created things, the reason they bend towards it so attentive an eye, is, that it speaks to their understanding some new evolution in the purposes of God—some new manifestation of his high attributes—some new and interesting step in the history of his sublime administration.

Now, we ought to be aware how it takes off, not from the intrinsic weight, but from the actual impression of our argument, that this devotedness to God which reigns in other places of the creation; this interest in him as the constant and essential principle of all enjoyment; this concern in the untaintedness of his glory; this delight in the survey of his perfections and his doings, are what the men of our corrupt and darkened world cannot sympathise with.

But however little we may enter into it, the Bible tells us by many intimations, that amongst those creatures who have not fallen from their allegiance, nor departed from the living God, God is their all—that love to him sits enthroned in their hearts, and fills them with all the ecstacy of an overwhelming affection—that a sense of grandeur never so elevates their souls, as when they look at the might and majesty of the Eternal—that no field of cloudless transparency so enchants them by the blissfulness of its visions, as when at the shrine of infinite and unspotted holiness, they bend themselves in raptured adoration—that no beauty so fascinates and attracts them, as does that moral beauty which throws a softening lustre over the awfulness of the Godhead—in a word, that the image of his character is ever present to their contemplations, and the unceasing joy of their sinless existence lies in the knowledge and the admiration of Deity.

Let us put forth an effort, and keep a steady hold of this consideration, for the deadness of our earthly imaginations makes an effort necessary; and we shall perceive, that though the world we live in, were the alone theatre of redemption, there is a something in the redemption itself that is fitted to draw the eye of an arrested universe towards it. Surely, surely, where delight in God is the constant enjoyment, and the earnest intelligent contemplation of God is the constant exercise, there is nothing in the whole compass of nature or of history, that can so set his adoring myriads upon the gaze, as some new and wondrous evolution of the character of God. Now this is found in the plan of our redemption; nor, do I see how in any transaction between the great Father of existence, and the children who have sprung from him, the moral attributes of the Deity could, if I may so express myself, be put to so severe and so delicate a test. It is true, that the great matters of sin and of salvation, fall without impression, on the heavy ears of a listless and alienated world. But they who, to use the language of the Bible, are light in the Lord, look otherwise at these things. They see sin in all its malignity, and salvation in all its mysterious greatness. Aye, and it would put them on the stretch of all their faculties, when they saw rebellion lifting up its standard against the Majesty of heaven, and the truth and the justice of God embarked on the threatenings he had uttered against all the doers of iniquity, and the honours of that august throne, which has the firm pillars of immutability to rest upon, linked with the fulfilment of the law that had come out from it; and when nothing else was looked for, but that God by putting forth the power of his wrath should accomplish his every denunciation, and vindicate the inflexibility of his government, and by one sweeping deed of vengeance, assert in the sight of all his creatures, the sovereignty which belonged to him—Oh! with what desire must they have pondered on his ways, when amid the urgency of all these demands which looked so high and so indispensable, they saw the unfoldings of the attribute of mercy—and how the Supreme Lawgiver was bending upon his guilty creatures an eye of tenderness—and how in his profound and unsearchable wisdom, he was devising for them some plan of restoration—and how the eternal Son had to move from his dwelling-place in heaven, to carry it forward through among all the difficulties by which it was encompassed—and how, after by the virtue of his mysterious sacrifice, he had magnified the glory of every other perfection, he made mercy rejoice over them all, and threw open a way by which we sinful and polluted wanderers, might, with the whole lustre of the Divine character untarnished, be re-admitted into fellowship with God, and be again brought back within the circle of his loyal and affectionate family.

Now, the essential character of such a transaction, viewed as a manifestation of God, does not hang upon the number of worlds, over which this sin and this salvation may have extended. We know that over this one world such an economy of wisdom and of mercy is instituted—and, even should this be the only world that is embraced by it, the moral display of the Godhead is mainly and substantially the same, as if it reached throughout the whole of that habitable extent which the science of astronomy has made known to us. By the disobedience of this one world, the law was trampled on—and, in the business of making truth and mercy to meet, and have a harmonious accomplishment on the men of this world, the dignity of God was put to the same trial; the justice of God appeared to lay the same immoveable barrier; the wisdom of God had to clear a way through the same difficulties; the forgiveness of God had to find the same mysterious conveyance to the sinners of a solitary world, as to the sinners of half a universe. The extent of the field upon which this question was decided, has no more influence on the question itself, than the figure or the dimensions of that field of combat, on which some great political question was fought, has on the importance or on the moral principles of the controversy that gave rise to it. This objection about the narrowness of the theatre, carries along with it all the grossness of materialism. To the eye of spiritual and intelligent beings, it is nothing. In their view, the redemption of a sinful world derives its chief interest from the display it gives of the mind and purposes of the Deity—and, should that world be but a single speck in the immensity of the works of God, the only way in which this affects their estimate of him is to magnify his loving kindness—who, rather than lose one solitary world of the myriads he has formed, would lavish all the riches of his beneficence and of his wisdom on the recovery of its guilty population.

Now, though it must be admitted that the Bible does not speak clearly or decisively as to the proper effect of redemption being extended to other worlds; it speaks most clearly and most decisively about the knowledge of it being disseminated amongst other orders of created intelligence than our own. But if the contemplation of God be their supreme enjoyment, then the very circumstance of our redemption being known to them, may invest it, even though it be but the redemption of one solitary world, with an importance as wide as the universe itself. It may spread amongst the hosts of immensity a new illustration of the character of him who is all their praise, and in looking toward whom every energy within them is moved to the exercise of a deep and delighted admiration. The scene of the transaction may be narrow in point of material extent; while in the transaction itself there may be such a moral dignity, as to blazon the perfections of the Godhead over the face of creation; and from the manifested glory of the Eternal, to send forth a tide of ecstacy, and of high gratulation, throughout the whole extent of his dependent provinces.

I will not, in proof of the position, that the history of our redemption is known in other and distant places of creation, and is matter of deep interest and feeling amongst other orders of created intelligence—I will not put down all the quotations which might be assembled together upon this argument. It is an impressive circumstance, that when Moses and Elias made a visit to our Saviour on the mount of transfiguration, and appeared in glory from heaven, the topic they brought along with them, and with which they were fraught, was the decease he was going to accomplish at Jerusalem. And however insipid the things of our salvation may be to an earthly understanding; we are made to know, that in the sufferings of Christ, and the glory which should follow, there is matter to attract the notice of celestial spirits, for these are the very things, says the Bible, which angels desire to look into. And however listlessly we, the dull and grovelling children of an exiled family, may feel about the perfections of the Godhead, and the display of those perfections in the economy of the Gospel; it is intimated to us in the book of God's message, that the creation has its districts and its provinces; and we accordingly read of thrones, and dominions, and principalities, and powers—and whether these terms denote the separate regions of government, or the beings who, by a commission granted from the sanctuary of heaven, sit in delegated authority over them—even in their eyes the mystery of Christ stands arrayed in all the splendour of unsearchable riches; for we are told that this mystery was revealed for the very intent, that unto the principalities and powers, in heavenly places, might be made known by the church, the manifold wisdom of God. And while we, whose prospect reaches not beyond the narrow limits of the corner we occupy, look on the dealings of God in the world, as carrying in them all the insignificancy of a provincial transaction; God himself, whose eye reaches to places which our eye hath not seen, nor our ear heard of, neither hath it entered into the imagination of our heart to conceive, stamps a universality on the whole matter of the Christian salvation, by such revelations as the following:—That he is to gather together in one all things in Christ, both which are in heaven, and which are in earth, even in him—and that at the name of Jesus every knee should bow, of things in heaven, and things in earth, and things under the earth—and that by him God reconciled all things unto himself, whether they be things in earth, or things in heaven.

We will not say in how far some of these passages extend the proper effect of that redemption which is by Christ Jesus, to other quarters of the universe of God; but they at least go to establish a widely disseminated knowledge of this transaction amongst the other orders of created intelligence. And they give us a distant glimpse of something more extended. They present a faint opening, through which may be seen some few traces of a wider and a nobler dispensation. They bring before us a dim transparency, on the other side of which the images of an obscure magnificence dazzle indistinctly upon the eye; and tell us, that in the economy of redemption, there is a grandeur commensurate to all that is known of the other works and purposes of the Eternal. They offer us no details; and man, who ought not to attempt a wisdom above that which is written, should never never put forth his hand to the drapery of that impenetrable curtain which God in his mysterious wisdom, has spread over those ways, of which it is but a very small portion that we know of them. But certain it is, that we know as much of them from the Bible; and the Infidel, with all the pride of his boasted astronomy, knows so little of them, from any power of observation, that the baseless argument of his, on which we have dwelt so long, is overborne in the light of all that positive evidence which God has poured around the record of his own testimony, and even in the light of its more obscure and casual intimations.

The minute and variegated details of the way in which this wondrous economy is extended, God has chosen to withhold from us; but he has oftener than once, made to us a broad and a general announcement of its dignity. He does not tell us whether the fountain opened in the house of Judah, for sin and for uncleanness, send forth its healing streams to other worlds than our own. He does not tell us the extent of the atonement. But he tells us that the atonement itself, known, as it is, among the myriads of the celestial, forms the high song of eternity; that the Lamb who was slain, is surrounded by the acclamations of one wide and universal empire; that the might of his wondrous achievements, spreads a tide of gratulation over the multitudes who are about his throne; and that there never ceases to ascend from the worshippers of him, who washed us from our sins in his blood, a voice loud as from numbers without number, sweet as from blessed voices uttering joy, when heaven rings jubilee, and loud hosannahs fill the eternal regions.

"And I beheld, and I heard the voice of many angels round about the throne, and the number of them was ten thousand times ten thousand, and thousands of thousands. Saying with a loud voice, Worthy is the Lamb that was slain, to receive power, and riches, and wisdom, and strength, and glory, and honour, and blessing. And every creature which is in heaven, and on the earth, and under the earth, and such as are in the sea, and all that are in them, heard I saying, Blessing, and honour, and glory, and power, be unto him that sitteth on the throne, and unto the Lamb, for ever and ever."

A king might have the whole of his reign crowded with the enterprises of glory; and by the might of his arms, and the wisdom of his counsels, might win the first reputation among the potentates of the world; and be idolized throughout all his provinces, for the wealth and the security that he had spread around them—and still it is conceivable, that by the act of a single day in behalf of a single family; by some soothing visitation of tenderness to a poor and solitary cottage; by some deed of compassion, which conferred enlargement and relief on one despairing sufferer; by some graceful movement of sensibility at a tale of wretchedness; by some noble effort of self-denial, in virtue of which he subdued his every purpose of revenge, and spread the mantle of a generous oblivion over the fault of the man who had insulted and aggrieved him; above all, by an exercise of pardon so skilfully administered, as that instead of bringing him down to a state of defencelessness against the provocation of future injuries, it threw a deeper sacredness over him, and stamped a more inviolable dignity than ever on his person and character:—why, my brethren, on the strength of one such performance done in a single hour, and reaching no farther in its immediate effects, than to one house, or to one individual, it is a most possible thing, that the highest monarch upon earth might draw such a lustre around him, as would eclipse the renown of all his public achievements—and that such a display of magnanimity, or of worth, beaming from the secrecy of his familiar moments, might waken a more cordial veneration in every bosom, than all the splendour of his conspicuous history—aye, and that it might pass down to posterity, as a more enduring monument of greatness, and raise him farther by its moral elevation, above the level of ordinary praise; and when he passes in review before the men of distant ages, may this deed of modest, gentle, unobtrusive virtue, be at all times appealed to, as the most sublime and touching memorial of his name.

In like manner, did the King eternal, immortal, and invisible, surrounded as he is with the splendours of a wide and everlasting monarchy, turn him to our humble habitation; and the footsteps of God manifest in the flesh, have been on the narrow spot of ground we occupy; and small though our mansion be, amid the orbs and the systems of immensity, hither hath the

King of glory bent his mysterious way, and entered the tabernacle of men, and in the disguise of a servant did he sojourn for years under the roof which canopies our obscure and solitary world. Yes, it is but a twinkling atom in the peopled infinity of worlds that are around it—but look to the moral grandeur of the transaction, and not to the material extent of the field upon which it was executed—and from the retirement of our dwelling-place, there may issue forth such a display of the Godhead, as will circulate the glories of his name amongst all his worshippers. Here sin entered. Here was the kind and universal beneficence of a Father, repaid by the ingratitude of a whole family. Here the law of God was dishonoured, and that too in the face of its proclaimed and unalterable sanctions. Here the mighty contest of the attributes was ended—and when justice put forth its demands, and truth called for the fulfilment of its warnings, and the immutability of God would not recede by a single iota, from any one of its positions, and all the severities he had ever uttered against the children of iniquity, seemed to gather into one cloud of threatening vengeance on the tenement that held us—did the visit of the only-begotten Son chase away all these obstacles to the triumph of mercy—and humble as the tenement may be, deeply shaded in the obscurity of insignificance as it is, among the statelier mansions which are on every side of it—yet will the recal of its exiled family never be forgotten—and the illustration that has been given here, of the mingled grace and majesty of God, will never lose its place among the themes and the acclamations of eternity.

And here it may be remarked, that as the earthly king who throws a moral aggrandisement around him, by the act of a single day, finds, that after its performance, he may have the space of many years for gathering to himself the triumphs of an extended reign—so the King who sits on high, and with whom one day is as a thousand years, and a thousand years as one day, will find, that after the period of that special administration is ended, by which this strayed world is again brought back within the limits of his favoured creation, there is room enough along the mighty track of eternity, for accumulating upon himself a glory as wide and as universal as is the extent of his dominions. You will allow the most illustrious of this world's potentates, to give some hour of his private history to a deed of cottage or of domestic tenderness; and every time you think of the interesting story, you will feel how sweetly and how gracefully the remembrance of it, blends itself with the fame of his public achievements. But still you think that there would not have been room enough for these achievements of his, had much of his time been spent, either amongst the habitations of the poor, or in the retirement of his own family; and you conceive, that it is because a single day bears so small a proportion to the time of his whole history, that he has been able to combine an interesting display of private worth, with all that brilliancy of exhibition, which has brought him down to posterity in the character of an august and a mighty sovereign.

Now apply this to the matter before us. Had the history of our redemption been confined within the limits of a single day, the argument that Infidelity has drawn from the multitude of other worlds, would never have been offered. It is true, that ours is but an insignificant portion of the territory of God—but if the attentions by which he has signalized it, had only taken up a single day, this would never have occurred to us as forming any sensible withdrawment of the mind of the Deity from the concerns of his vast and universal government. It is the time which the plan of our salvation requires, that startles all those on whom this argument has any impression. It is the time taken up about this paltry world, which they feel to be out of proportion to the number of other worlds, and to the immensity of the surrounding creation. Now, to meet this impression, I do not insist at present on what I have already brought forward, that God, whose ways are not as our ways, can have his eye at the same instant on every place, and can divide and diversify his attention into any number of distinct exercises. What I have now to remark, is, that the Infidel who urges the astronomical objection to the truth of Christianity, is only looking with half an eye to the principle on which it rests. Carry out the principle, and the objection vanishes. He looks abroad on the immensity of space, and tells us how impossible it is, that this narrow corner of it can be so distinguished by the attentions of the Deity. Why does he not also look abroad on the magnificence of eternity; and perceive how the whole period of these peculiar attentions, how the whole time which elapses between the fall of man and the consummation of the scheme of his recovery, is but the twinkling of a moment to the mighty roll of innumerable ages? The whole interval between the time of Jesus Christ's leaving his Father's abode, to sojourn amongst us, to that time when he shall have put all his enemies under his feet, and delivered up the kingdom to God even his Father, that God may be all in all; the whole of this interval bears as small a proportion to the whole of the Almighty's reign, as this solitary world does to the universe around it, and an infinitely smaller proportion than any time, however short, which an earthly monarch spends on some enterprise of private benevolence, does to the whole walk of his public and recorded history.

Why then does not the man, who can shoot his conceptions so sublimely abroad over the field of an immensity that knows no limits—why does he not also shoot them forward through the vista of a succession, that ever flows without stop and without termination? He has burst across the confines of this world's habitation in space, and out of the field which lies on the other side of it, has he gathered an argument against the truth of revelation. I feel that I have nothing to do but to burst across the confines of this world's history in time, and out of the futurity which lies beyond it, can I gather that which will blow the argument to pieces, or stamp upon it all the narrowness of a partial and mistaken calculation. The day is coming, when the whole of this wondrous history shall be looked back upon by the eye of remembrance, and be regarded as one incident in the extended annals of creation, and with all the illustration and all the glory it has thrown on the character of the Deity, will it be seen as a single step in the evolution of his designs; and long as the time may appear, from the first act of our redemption to its final accomplishment, and close and exclusive as we may think the attentions of God upon it, it will be found that it has left him room enough for all his concerns, and that on the high scale of eternity, it is but one of those passing and ephemeral transactions, which crowd the history of a never-ending administration.

DISCOURSE V.

ON THE SYMPATHY THAT IS FELT FOR MAN IN THE DISTANT PLACES OF CREATION.

"I say unto you, that likewise joy shall be in heaven over one sinner that repenteth, more than over ninety and nine just persons which need no repentance."

LUKE xv. 7.

I HAVE already attempted at full length to establish the position, that the infidel argument of astronomers goes to expunge a natural perfection from the character of God, even that wondrous property of his, by which he, at the same instant of time, can bend a close and a careful attention on a countless diversity of objects, and diffuse the intimacy of his power and of his presence, from the greatest to the minutest and most insignificant of them all. I also adverted shortly to this other circumstance, that it went to impair a moral attribute of the Deity. It goes to impair the benevolence of his nature. It is saying much for the benevolence of God, to say, that a single world, or a single system, is not enough for it—that it must have the spread of a mightier region, on which it may pour forth a tide of exuberancy throughout all its provinces—that as far as our vision can carry us, it has strewed immensity with the floating receptacles of life, and has stretched over each of them the garniture of such a sky as mantles our own habitation—and that even from distances which are far beyond the reach of human eye, the songs of gratitude and praise may now be arising to the one God, who sits surrounded by the regards of his one great and universal family.

Now it is saying much for the benevolence of God, to say that it sends forth these wide and distant emanations over the surface of a territory so ample, that the world we inhabit, lying imbedded as it does, amidst so much surrounding greatness, shrinks into a point that to the universal eye might appear to be almost imperceptible. But does it not add to the power and to the perfection of this universal eye, that at the very moment it is taking a comprehensive survey of the vast, it can fasten a steady and undistracted attention on each minute and separate portion of it; that at the very moment it is looking at all worlds, it can look most pointedly and most intelligently to each of them; that at the very moment it sweeps the field of immensity, it can settle all the earnestness of its regards upon every distinct handbreadth of that field; that at the very moment at which it embraces the totality of existence, it can send a most thorough and penetrating inspection into each of its details, and into every one of its endless diversities? You cannot fail to perceive how much this adds to the power of the all-seeing eye. Tell me then, if it do not add as much perfection to the benevolence of God, that while it is expatiating over the vast field of created things, there is not one portion of the field overlooked by it; that while it scatters blessings over the whole of an infinite range, it causes them to descend in a shower of plenty on every separate habitation; that while his arm is underneath and round about all worlds, he enters within the precincts of every one of them, and gives a care and a tenderness to each individual of their teeming population. Oh! does not the God, who is said to be love, shed over this attribute of his, its finest illustration! when while he sits in the highest heaven, and pours out his fulness on the whole subordinate domain of nature and of providence, he bows a pitying regard on the very humblest of his children, and sends his reviving Spirit into every heart, and cheers by his presence every home, and provides for the wants of every family, and watches every sick-bed, and listens to the complaints of every sufferer; and while by his wondrous mind the weight of universal government is borne, oh! is it not more wondrous and more excellent still, that he feels for every sorrow, and has an ear open to every prayer.

" It doth not yet appear what we shall be," says the apostle John, " but we know that when he shall appear, we shall be like him, for we shall see him as he is." It is the present lot of the angels, that they behold the face of our Father in heaven, and it would seem as if the effect of this was to form and to perpetuate in them the moral likeness of himself, and that they reflect back upon him his own image, and that thus a diffused resemblance to the Godhead, is kept up amongst all those adoring worshippers who live in the near and rejoicing contemplation of the Godhead. Mark then how that peculiar and endearing feature in the goodness of the Deity, which we have just now adverted to—mark how beauteously it is reflected downwards upon us in the revealed attitude of angels. From the high eminences of heaven, are they bending a wakeful regard over the men of this sinful world; and the repentance of every one of them spreads a joy and a high gratulation throughout all its dwelling places. Put this trait of the angelic character into contrast with the dark and louring spirit of an infidel. He is told of the multitude of other worlds, and he feels a kindling magnificence in the conception, and he is seduced by an elevation which he cannot carry, and from this airy summit does he look down on the insignificance of the world we occupy, and pronounces it to be unworthy of those visits and of those attentions which we read of in the New Testament. He is unable to wing his upward way along the scale, either of moral or of natural perfection; and when the wonderful extent of the field is made known to him, over which the wealth of the Divinity is lavished—there he stops, and wilders, and altogether misses this essential perception, that the power and perfection of the Divinity are not more displayed by the mere magnitude of the field, than they are by that minute and exquisite filling up, which leaves not its smallest portions neglected; but which imprints the fulness of the Godhead upon every one of them; and proves, by every flower of the pathless desert, as well as by every orb of immensity, how this unsearchable Being can care for all, and provide for all, and throned in mystery too high for us, can, throughout every instant of time, keep his attentive eye on every separate thing that he has formed, and by an act of his thoughtful and presiding intelligence, can constantly embrace all.

But God, compassed about as he is with light inaccessible, and full of glory, lies so hidden from the ken and conception of all our faculties, that the spirit of man sinks exhausted by its attempts to comprehend him. Could the image of the Supreme be placed direct before the eye of the mind, that flood of splendour, which is ever issuing from him on all who have the privilege of beholding, would not only dazzle, but overpower us. And, therefore it is, that I bid you look to the reflection of that image, and thus to take a view of its mitigated glories, and to gather the lineaments of the Godhead in the face of those righteous angels, who have never thrown away from them the resemblance in which they were created; and, unable as you are to support the grace and the majesty of that countenance, before which the sons and the prophets of other days fell, and became as dead men, let us, before we bring this argument to a close, borrow one lesson of him who sitteth on the throne, from the aspect and the revealed doings of those who are surrounding it.

The infidel, then, as he widens the field of his contemplations, would suffer its every separate object to die away into forgetfulness: these angels, expatiating as they do, over the range of a loftier universality, are represented as all awake to the history of each of its distinct and subordinate provinces. The infidel, with his mind afloat among suns and among systems, can find no place in his already occupied regards, for that humble planet which lodges and accommodates our species: the angels, standing on a loftier summit, and with a mightier prospect of creation before them, are yet represented as looking down on this single world, and attentively marking the every feeling and the every demand of all its families. The infidel, by sinking us down to an unnoticeable minuteness, would lose sight of our dwelling-place altogether, and spread a darkening shroud of oblivion over all the concerns and all the interests of men: but the angels will not so abandon us; and undazzled by the whole surpassing grandeur of that scenery which is around them, are they revealed as directing all the fulness of their regard to this our habitation, and casting a longing and a benignant eye on ourselves and on our children. The infidel will tell us of those worlds which roll afar, and the number of which outstrips the arithmetic of the human understanding—and then with the hardness of an unfeeling calculation, will he consign the one we occupy, with all its guilty generations, to despair. But he who counts the number of the stars, is set forth to us as looking at every inhabitant among the millions of our species, and by the word of the Gospel beckoning to him with the hand of invitation, and on the very first step of his return, as moving towards him with all the eagerness of the prodigal's father, to receive him back again into that presence from which he had wandered. And as to this world, in favour of which the scowling infidel will not permit one solitary movement, all heaven is represented as in a stir about its restoration; and there cannot a single son, or a single daughter, be recalled from sin unto righteousness, without an acclamation of joy amongst the hosts of Paradise. Aye, and I can say it of the humblest and the unworthiest of you all, that the eye of angels is upon him, and that his repentance would, at this moment, send forth a wave of delighted sensibility throughout the mighty throng of their innumerable legions.

Now, the single question I have to ask, is, On which of the two sides of this contrast do we see most of the impress of heaven? Which of the two would be most glorifying to God? Which of them carries upon it most of that evidence which lies in its having a celestial character? For if it be the side of the infidel, then must all our hopes expire with the ratifying of that fatal sentence, by which the world is doomed, through its insignificancy, to perpetual exclusion from the attentions of the Godhead. I have long been knocking at the door of your understanding, and have tried to find an admittance to it for many an argument. I now make my appeal to the sensibilities of your heart; and tell me, to whom does the moral feeling within it yield its readiest testimony—to the infidel, who would make this world of ours vanish away into abandonment—or to those angels, who ring throughout all their mansions the hosannas of joy, over every one individual of its repentant population?

And here I cannot omit to take advantage of that opening with which our Saviour has furnished us, by the parables of this chapter, and admits us into a familiar view of that principle on which the inhabitants of heaven are so awake to the deliverance and the restoration of our species. To illustrate the difference in the reach of knowledge and of affection, between a man and an angel, let us think of the difference of reach between one man and another. You may often witness a man, who feels neither tenderness nor care beyond the precincts of his own family; but who, on the strength of those instinctive fondnesses which nature has implanted in his bosom, may earn the character of an amiable father, or a kind husband, or a bright example of all that is soft and endearing in the relations of domestic society. Now, conceive him, in addition to all this, to carry his affections abroad, without, at the same time, any abatement of their intensity towards the objects which are at home—that stepping across the limits of the house he occupies, he takes an interest in the families which are near him—that he lends his services to the town or the district wherein he is placed, and gives up a portion of his time to the thoughtful labours of a humane and public-spirited citizen. By this enlargement in the sphere of his attention, he has extended his reach; and, provided he has not done so at the expense of that regard which is due to his family, a thing which, cramped and confined as we are, we are very apt, in the exercise of our humble faculties, to do—I put it to you, whether by extending the reach of his views and his affections, he has not extended his worth and his moral respectability along with it?

But I can conceive a still farther enlargement. I can figure to myself a man, whose wakeful sympathy overflows the field of his own immediate neighbourhood—to whom the name of country comes with all the omnipotence of a charm upon his heart, and with all the urgency of a most righteous and resistless claim upon his services—who never hears the name of Britain sounded in his ears, but it stirs up all his enthusiasm in behalf of the worth and the welfare of its people—who gives himself up, with all the devotedness of a passion, to the best and the purest objects of patriotism—and who, spurning away from him the vulgarities of party ambition, separates his life and his labours to the fine pursuit of augmenting the science, or the virtue, or the substantial prosperity of his nation. Oh! could such a man retain all the tenderness, and fulfil all the duties which home and which neighbourhood require of him, and at the same time, expatiate in the might of his untired faculties, on so wide a field of benevolent contemplation—would not this extension of reach place him still higher than before, on the scale both of moral and intellectual gradation, and give him a still brighter and more enduring name in the records of human excellence?

And, lastly, I can conceive a still loftier flight of humanity—a man, the aspiring of whose heart for the good of man, knows no limitations—whose longings, and whose conceptions on this subject, overleap all the barriers of geography—who, looking on himself as a brother of the species, links every spare energy which belongs to him, with the cause of its amelioration—who can embrace within the grasp of his ample desires, the whole family of mankind—and who, in obedience to a heaven-born movement of principle within him, separates himself to some big and busy enterprise, which is to tell on the moral destinies of the world. Oh! could such a man mix up the softenings of private virtue, with the habit of so sublime a comprehension—if, amid those magnificent darings of thought and of performance, the mildness of his benignant eye could still continue to cheer the retreat of his family, and to spread the charm and the sacredness of piety among all its members—could he even mingle himself in all the gentleness of a soothed and a smiling heart, with the playfulness of his children—and also find strength to shed the blessings of his presence and his counsel over the vicinity around him;—oh! would not the combination of so much grace with so much loftiness, only serve the more to aggrandize him? Would not the one ingredient of a character so rare, go to illustrate and to magnify the other? And would not you pronounce him to be the fairest specimen of our nature, who could so call out all your tenderness, while he challenged and compelled all your veneration?

Nor can I proceed, at this point of my argument, without adverting to the way in which this last and this largest style of benevolence is exemplified in our own country—where the spirit of the Gospel has given to many of its enlightened disciples, the impulse of such a philanthropy, as carries abroad their wishes and their endeavours to the very outskirts of human population—a philanthropy, of which, if you asked the extent or the boundary of its field, we should answer, in the language of inspiration, that the field is the world—a philanthropy, which overlooks all the distinctions of cast and of colour, and spreads its ample regards over the whole brotherhood of the species—a philanthropy, which attaches itself to man in the general; to man throughout all his varieties; to man as the partaker of one common nature, and who, in whatever clime or latitude you may meet with him, is found to breathe the same sympathies, and to possess the same high capabilities both of bliss and of improvement. It is true, that, upon this subject, there is often a loose and unsettled magnificence of thought, which is fruitful of nothing but empty speculation. But, the men to whom I allude, have not imaged the enterprise in the form of a thing unknown. They have given it a local habitation. They have bodied it forth in deed and in accomplishment. They have turned the dream into a reality. In them, the power of a lofty generalization meets with its happiest attemperment, in the principle and perseverance, and all the chastening and subduing virtues of the New Testament. And, were I in search of that fine union of grace and of greatness which I have now been insisting on, and, in virtue of which, the enlightened Christian can at once find room in his bosom for the concerns of universal humanity, and for the play of kindliness towards every individual he meets with—I could no where more readily expect to find it, than with the worthies of our own land—the Howard of a former generation, who paced it over Europe in quest of the unseen wretchedness which abounds in it—or in such men of our present generation, as Wilberforce, who lifted his unwearied voice against the biggest outrage ever practised on our nature, till he wrought its extermination—and Clarkson, who plied his assiduous task at rearing the materials of its impressive history, and, at length carried, for this righteous cause, the mind of Parliament—and Carey, from whose hand the generations of the East are now receiving the elements of their moral renovation—and, in fine, those holy and devoted men, who count not their lives dear unto them; but, going forth every year from the island of our habitation, carry the message of heaven over the face of the world; and, in the front of severest obloquy, are now labouring in remotest lands; and are reclaiming another and another portion from the wastes of dark and fallen humanity; and are widening the domains of gospel light and gospel principle amongst them; and are spreading a moral beauty around the every spot on which they pitch their lowly tabernacle; and are at length compelling even the eye and the testimony of gainsayers, by the success of their noble enterprise; and are forcing the exclamation of delighted surprise from the charmed and the arrested traveller, as he looks at the softening tints which they are now spreading over the wilderness, and as he hears the sound of the chapel bell, and as in those haunts where, at the distance of half a generation, savages would have scowled upon his path, he regales himself with the hum of missionary schools, and the lovely spectacle of peaceful and Christian villages.

Such, then, is the benevolence, at once so gentle and so lofty, of those men, who, sanctified by the faith that is in Jesus, have had their hearts visited from heaven by a beam of warmth and of sacredness. What, then, I should like to know, is the benevolence of the place from whence such an influence cometh? How wide is the compass of this virtue there, and how exquisite is the feeling of its tenderness, and how pure and how fervent are its aspirings among those unfallen beings who have no darkness, and no encumbering weight of corruption to strive against? Angels have a mightier reach of contemplation. Angels can look upon this world and all which it inherits, as the part of a larger family. Angels were in the full exercise of their powers even at the first infancy of our species, and shared in the gratulations of that period, when at the birth of humanity all intelligent nature felt a gladdening impulse, and the morning stars sang together for joy. They loved us even with the love which a family on earth bears to a younger sister; and the very childhood of our tinier faculties did only serve the more to endear us to them; and though born at a later hour in the history of creation, did they regard us as heirs of the same destiny with themselves, to rise along with them in the scale of moral elevation, to bow at the same footstool, and to partake in those high dispensations of a parent's kindness and a parent's care, which are ever emanating from the throne of the Eternal on all the members of a duteous and affectionate family. Take the reach of an angel's mind, but, at the same time, take the seraphic fervour of an angel's benevolence along with it; how, from the eminence on which he stands he may have an eye upon many worlds, and a remembrance upon the origin and the successive concerns of every one of them; how he may feel the full force of a most affecting relationship with the habitants of each, as the offspring of one common Father; and though it be both the effect and the evidence of our depravity, that we cannot sympathise with these pure and generous ardours of a celestial spirit; how it may consist with the lofty comprehension, and the ever-breathing love of an angel, that he can both shoot his benevolence abroad over a mighty expanse of planets and of systems, and lavish a flood of tenderness on each individual of their teeming population.

Keep all this in view, and you cannot fail to perceive how the principle so finely and so copiously illustrated in this chapter, may be brought to meet the infidelity we have thus long been employed in combating. It was nature, and the experience of every bosom will affirm it—it was nature in the shepherd to leave the ninety and nine of his flock forgotten and alone in the wilderness, and betaking himself to the mountains, to give all his labour and all his concern to the pursuit of one solitary wanderer. It was nature; and we are told in the passage before us, that it is such a portion of nature as belongs not merely to men, but to angels; when the woman, with her mind in a state of listlessness as to the nine pieces of silver that were in secure custody, turned the whole force of her anxiety to the one piece which she had lost, and for which she had to light a candle, and to sweep the house, and to search diligently until she found it. It was nature in her to rejoice more over that piece, than over all the rest of them, and to tell it abroad among friends and neighbours, that they might rejoice along with her—aye, and sadly effaced as humanity is, in all her original lineaments, this is a part of our nature, the very movements of which are experienced in heaven, "where there is more joy over one sinner that repenteth, than over ninety and nine just persons who need no repentance." For any thing I know, the every planet that rolls in the immensity around me, may be a land of righteousness; and be a member of the household of God; and have her secure dwelling-place within that ample limit, which embraces his great and universal family. But I know at least of one wanderer; and how woefully she has strayed from peace and from purity; and how in dreary alienation from him who made her, she has bewildered herself amongst those many devious tracks, which have carried her afar from the path of immortality; and how sadly tarnished all those beauties and felicities are, which promised, on that morning of her existence when God looked on her, and saw that all was very good—which promised so richly to bless and to adorn her; and how in the eye of the whole unfallen creation, she has renounced all this goodliness, and is fast departing away from them into guilt, and wretchedness, and shame. Oh! if there be any truth in this chapter, and any sweet or touching nature in the principle which runs throughout all its parables, let us cease to wonder, though they who surround the throne of love should be looking so intently towards us—or though, in the way by which they have singled us out, all the other orbs of space should, for one short season, on the scale of eternity, appear to be forgotten—or though, for every step of her recovery, and for every individual who is rendered back again to the fold from which he was separated, another and another message of triumph should be made to circulate amongst the hosts of paradise—or though, lost as we are, and sunk in depravity as we are, all the sympathies of heaven should now be awake on the enterprise of him who has travailed, in the greatness of his strength, to seek and to save us.

And here I cannot but remark how fine a harmony there is between the law of sympathetic nature in heaven, and the most touching exhibitions of it on the face of our world. When one of a numerous household droops under the power of disease, is not that the one to whom all the tenderness is turned, and who, in a manner, monopolises the inquiries of his neighbourhood, and the care of his family? When the sighing of the midnight storm sends a dismal foreboding into the mother's heart, to whom of all her offspring, I would ask, are her thoughts and her anxieties then wandering? Is it not to her sailor boy whom her fancy has placed amid the rude and angry surges of the ocean? Does not this, the hour of his apprehended danger, concentrate upon him the whole force of her wakeful meditations? And does not he engross, for a season, her every sensibility, and her every prayer? We sometimes hear of shipwrecked passengers thrown upon a barbarous shore; and seized upon by its prowling inhabitants; and hurried away through the tracks of a dreary and unknown wilderness; and sold into captivity; and loaded with the fetters of irrecoverable bondage; and who stripped of every other liberty but the liberty of thought, feel even this to be another ingredient of wretchedness, for what can they think of but home, and as all its kind and tender imagery comes upon their remembrance, how can they think of it but in the bitterness of despair? Oh tell me when the fame of all this disaster reaches his family, who is the member of it to whom is directed the full tide of its griefs and of its sympathies? Who is it that, for weeks and for months, usurps their every feeling, and calls out their largest sacrifices, and sets them to the busiest expedients for getting him back again?

Who is it that makes them forgetful of themselves and of all around them; and tell me if you can assign a limit to the pains, and the exertions, and the surrenders which afflicted parents and weeping sisters would make to seek and to save him?

Now conceive, as we are warranted to do by the parables of this chapter, the principle of all these earthly exhibitions to be in full operation around the throne of God. Conceive the universe to be one secure and rejoicing family, and that this alienated world is the only strayed, or only captive member belonging to it; and we shall cease to wonder, that from the first period of the captivity of our species, down to the consummation of their history in time, there should be such a movement in heaven; or that angels should so often have sped their commissioned way on the errand of our recovery; or that the Son of God should have bowed himself down to the burden of our mysterious atonement; or that the Spirit of God should now, by the busy variety of his all-powerful influences, be carrying forward that dispensation of grace which is to make us meet for re-admittance into the mansions of the celestial. Only think of love as the reigning principle there; of love, as sending forth its energies and aspirations to the quarter where its object is most in danger of being for ever lost to it; of love, as called forth by this single circumstance to its uttermost exertion, and the most exquisite feeling of its tenderness; and then shall we come to a distinct and familiar explanation of this whole mystery: Nor shall we resist by our incredulity, the gospel message any longer, though it tells us, that throughout the whole of this world's history, long in our eyes, but only a little month in the high periods of immortality, so much of the vigilance, and so much of the earnestness of heaven, should have been expended on the recovery of its guilty population.

There is another touching trait of nature, which goes finely to heighten this principle, and still more forcibly to demonstrate its application to our present argument. So long as the dying child of David was alive, he was kept on the stretch of anxiety and of suffering with regard to it. When it expired, he arose and comforted himself. This narrative of King David is in harmony with all that we experience of our own movements and our own sensibilities. It is the power of uncertainty which gives them so active and so interesting a play in our bosoms; and which heightens all our regards to a tenfold pitch of feeling and of exercise; and which fixes down our watchfulness upon our infant's dying bed; and which keeps us so painfully alive to every turn and to every symptom in the progress of its malady; and which draws out all our affections for it to a degree of intensity that is quite unutterable; and which urges us on to ply our every effort and our every expedient, till hope withdraw its lingering beam, or till death shut the eyes of our beloved in the slumber of its long and its last repose.

I know not who of you have your names written in the book of life—nor can I tell if this be known to the angels which are in heaven. While in the land of living men, you are under the power and application of a remedy, which if taken as the Gospel prescribes, will renovate the soul, and altogether prepare it for the bloom and the vigour of immortality. Wonder not then that with this principle of uncertainty in such full operation, ministers should feel for you: or angels should feel for you; or all the sensibilities of heaven should be awake upon the symptoms of your grace and reformation; or the eyes of those who stand upon the high eminences of the celestial world, should be so earnestly fixed on the every footstep and new evolution of your moral history. Such a consideration as this should do something more than silence the infidel objection. It should give a practical effect to the calls of repentance. How will it go to aggravate the whole guilt of our impenitency, should we stand out against the power and the tenderness of these manifold applications—the voice of a beseeching God upon us—the word of salvation at our very door—the free offer of strength and of acceptance sounded in our hearing—the Spirit in readiness with his agency to meet our every desire and our every inquiry—angels beckoning us to their company—and the very first movements of our awakened conscience, drawing upon us all their regards and all their earnestness.

DISCOURSE VI.

ON THE CONTEST FOR AN ASCENDENCY OVER MAN, AMONGST THE HIGHER ORDERS OF INTELLIGENCE.

"And having spoiled principalities and powers, he made a show of them openly, triumphing over them in it."

Col. ii. 15.

Though these Astronomical Discourses be now drawing to a close, it is not because I feel that much more might not be said on the subject of them, both in the way of argument and of illustration. The whole of the Infidel difficulty proceeds upon the assumption, that the exclusive bearing of Christianity is upon the people of our earth; that this solitary planet is in no way implicated with the concerns of a wider dispensation; that the revelation we have of the dealings of God, in this district of his empire, does not suit and subordinate itself to a system of moral administration, as extended as is the whole of his monarchy. Or, in other words, because Infidels have not access to the whole truth, will they refuse a part of it, however well attested or well accredited it may be; because a mantle of deep obscurity rests on the government of God, when taken in all its eternity and all its entireness, will they shut their eyes against that allowance of light which has been made to pass downwards upon our world from time to time, through so many partial unfoldings; and till they are made to know the share which other planets have in these communications of mercy, will they turn them away from the actual message which has come to their own door, and will neither examine its credentials, nor be alarmed by its warnings, nor be won by the tenderness of its invitations.

On that day when the secrets of all hearts shall be revealed, there will be found such a wilful duplicity and darkening of the mind in the whole of this proceeding, as shall bring down upon it the burden of a righteous condemnation. But, even now, does it lie open to the rebuke of philosophy, when the soundness and the consistency of her principles are brought faithfully to bear upon it. Were the character of modern science rightly understood, it would be seen, that the very thing which gave such strength and sureness to all her conclusions, was that humility of spirit which belonged to her. She promulgates all that is positively known; but she maintains the strictest silence and modesty about all that is unknown. She thankfully accepts of evidence wherever it can be found; nor does she spurn away from her the very humblest contribution of such doctrine, as can be witnessed by human observation, or can be attested by human veracity. But with all this she can hold out most sternly against that power of eloquence and fancy, which often throws so bewitching a charm over the plausibilities of ingenious speculation. Truth is the alone idol of her reverence; and did she at all times keep by her attachments, nor throw them away when theology submitted to her cognizance its demonstrations and its claims, we should not despair of witnessing as great a revolution in those prevailing habitudes of thought which obtain throughout our literary establishments, on the subject of Christianity, as that which has actually taken place in the philosophy of external nature. This is the first field on which have been successfully practised the experimental lessons of Bacon; and they who are conversant with these matters, know how great and how general a uniformity of doctrine now prevails in the sciences of astronomy, and mechanics, and chemistry, and almost all the other departments in the history and philosophy of matter. But this uniformity stands strikingly contrasted with the diversity of our moral systems, with the restless fluctuations both of language and of sentiment which are taking place in the philosophy of mind, with the palpable fact, that every new course of instruction upon this subject, has some new articles, or some new explanations to peculiarise it: and all this is to be attributed, not to the progress of the science, not to a growing, but to an alternating movement, not to its perpetual additions, but to its perpetual vibrations.

I mean not to assert the futility of moral

science, or to deny her importance, or to insist on the utter hopelessness of her advancement. The Baconian method will not probably push forward her discoveries with such a rapidity, or to such an extent, as many of her sanguine disciples have anticipated. But if the spirit and the maxims of this philosophy were at all times proceeded upon, it would certainly check that rashness and variety of excogitation, in virtue of which it may almost be said, that every new course presents us with a new system, and that every new teacher has some singularity or other to characterise him. She may be able to make out an exact transcript of the phenomena of mind, and in so doing, she yields a most important contribution to the stock of human acquirements. But when she attempts to grope her darkling way through the counsels of the Deity, and the futurities of his administration; when, without one passing acknowledgement to the embassy which professes to have come from him, or to the facts and to the testimonies by which it has so illustriously been vindicated, she launches forth her own speculations on the character of God, and the destiny of man;

when, though this be a subject on which neither the recollections of history, nor the ephemeral experience of any single life, can furnish one observation to enlighten her, she will nevertheless utter her own plausibilities, not merely with a contemptuous neglect of the Bible, but in direct opposition to it; then it is high time to remind her of the difference between the reverie of him who has not seen God, and the well-accredited declaration of him who was in the beginning with God, and was God; and to tell her, that this so far from being the argument of an ignoble fanaticism, is in harmony with the very argument upon which the science of experiment has been reared, and by which it has been at length delivered from the influence of theory, and purified of all its vain and visionary splendours.

In my last Discourses, I have attempted to collect from the records of God's actual communication to the world, such traces of relationship between other orders of being and the great family of mankind, as serve to prove that Christianity is not so paltry and provincial a system as Infidelity presumes it to be. And as I said before, I have not exhausted all that may legitimately be derived upon this subject from the informations of Scripture. I have adverted, it is true, to the knowledge of our moral history, which obtains throughout other provinces of the intelligent creation. I have asserted the universal importance which this may confer on the transactions even of one planet, in as much as it may spread an honourable display of the Godhead amongst all the mansions of infinity. I have attempted to expatiate on the argument, that an event little in itself, may be so pregnant with character, as to furnish all the worshippers of heaven with a theme of praise for eternity. I have stated that nothing is of magnitude in their eyes, but that which serves to endear to them the Father of their spirits, or to shed a lustre over the glory of his incomprehensible attributes—and that thus, from the redemption even of our solitary species, there may go forth such an exhibition of the Deity, as shall bear the triumphs of his name to the very outskirts of the universe.

I have farther adverted to another distinct Scriptural intimation, that the state of fallen man was not only matter of knowledge to other orders of creation, but was also matter of deep regret and affectionate sympathy; that, agreeably to such laws of sympathy as are most familiar even to human observation, the very wretchedness of our condition was fitted to concentrate upon us the feelings, and the attentions, and the services, of the celestial—to single us out for a time to the gaze of their most earnest and unceasing contemplation—to draw forth all that was kind and all that was tender within them—and just in proportion to the need and to the helplessness of us miserable exiles from the family of God, to multiply upon us the regards, and call out in our behalf the fond and eager exertions of those who had never wandered away from him. This appears from the Bible to be the style of that benevolence which glows and which circulates around the throne of heaven. It is the very benevolence which emanates from the throne itself, and the attentions of which have for so many thousand years signalized the inhabitants of our world. This may look a long period for so paltry a world. But how have Infidels come to their conception that our world is so paltry? By looking abroad over the countless systems of immensity. But why then have they missed the conception, that the time of those peculiar visitations, which they look upon as so disproportionate to the magnitude of this earth, is just as evanescent as the earth itself is insignificant? Why look they not abroad on the countless generations of eternity; and thus come back to the conclusion, that after all, the redemption of our species is but an ephemeral doing in the history of intelligent nature; that it leaves the Author of it room for all the accomplishments of a wise and equal administration; and not to mention, that even during the progress of it, it withdraws not a single thought or a single energy of his, from other fields of creation, that there remains time enough to him for carrying round the visitations of as striking and as peculiar a tenderness, over the whole extent of his great and universal monarchy?

It might serve still farther to incorporate the

concerns of our planet with the general history of moral and intelligent beings, to state, not merely the knowledge which they take of us, and not merely the compassionate anxiety which they feel for us; but to state the importance derived to our world from its being the actual theatre of a keen and ambitious contest amongst the upper orders of creation. You know that how for the possession of a very small and insulated territory, the mightiest empires of the world, have put forth all their resources; and on some field of mustering competition, have monarchs met, and embarked for victory, all the pride of a country's talent, and all the flower and strength of a country's population. The solitary island around which so many fleets are hovering, and on the shores of which so many armed men are descending, as to an arena of hostility, may well wonder at its own unlooked for estimation. But other principles are animating the battle; and the glory of nations is at stake; and a much higher result is in the contemplation of each party, than the gain of so humble an acquirement as the primary object of the war; and honour, dearer to many a bosom than existence, is now the interest on which so much blood and so much treasure is expended; and the stirring spirit of emulation has now got hold of the combatants; and thus, amid all the insignificancy which attaches to the material origin of the contest, do both the eagerness and the extent of it, receive from the constitution of our nature, their most full and adequate explanation.

Now, if this be also the principle of higher natures—if, on the one hand, God be jealous of his honour, and on the other, there be proud and exalted spirits, who scowl defiance at him and at his monarchy—if, on the side of heaven, there be an angelic host rallying around the standard of loyalty, who flee with alacrity at the bidding of the Almighty, who are devoted to his glory, and feel a rejoicing interest in the evolution of his counsels; and if, on the side of hell, there be a sullen front of resistance, a hate and malice inextinguishable, an unquelled daring of revenge to baffle the wisdom of the Eternal, and to arrest the hand, and to defeat the purposes of Omnipotence—then let the material prize of victory be insignificant as it may, it is the victory in itself which upholds the impulse of this keen and stimulated rivalry. If, by the sagacity of one infernal mind, a single planet has been seduced from its allegiance, and been brought under the ascendancy of him who is called in Scripture, "the god of this world," and if the errand on which our Redeemer came, was to destroy the works of the devil—then let this planet have all the littleness which astronomy has assigned to it—call it what it is, one of the smaller islets which float on the ocean of vacancy; it has become the theatre of such a competition, as may have all the desires and all the energies of a divided universe embarked upon it. It involves in it other objects than the single recovery of our species. It decides higher questions. It stands linked with the supremacy of God, and will at length demonstrate the way in which he inflicts chastisement and overthrow upon all his enemies. I know not if our rebellious world be the only stronghold which Satan is possessed of, or if it be but the single post of an extended warfare, that is now going on between the powers of light and of darkness. But be it the one or the other, the parties are in array, and the spirit of the contest is in full energy, and the honour of mighty combatants is at stake; and let us therefore cease to wonder that our humble residence has been made the theatre of so busy an operation, or that the ambition of loftier natures has here put forth all its desire and all its strenuousness.

This unfolds to us another of those high and extensive bearings, which the moral history of our globe may have on the system of God's universal administration. Were an enemy to touch the shore of this high-minded country, and to occupy so much as one of the humblest of its villages, and there to seduce the natives from their loyalty, and to sit down along with them in entrenched defiance to all the threats, and to all the preparations of an insulted empire—oh! how would the cry of wounded pride resound throughout all the ranks and varieties of our mighty population; and this very movement of indignancy would reach the king upon his throne; and circulate among those who stood in all the grandeur of chieftainship around him; and be heard to thrill in the eloquence of Parliament; and spread so resistless an appeal to a nation's honour, and a nation's patriotism, that the trumpet of war would summon to its call, all the spirit and all the willing energies of our kingdom; and rather than sit down in patient endurance under the burning disgrace of such a violation, would the whole of its strength and resources be embarked upon the contest; and never never would we let down our exertions and our sacrifices, till either our deluded countrymen were reclaimed, or till the whole of this offence, were, by one righteous act of vengeance, swept away altogether from the face of the territory it deformed.

The Bible is always most full and most explanatory on those points of revelation in which men are personally interested. But it does at times offer a dim transparency, through which may be caught a partial view of such designs and of such enterprises as are now afloat among the upper orders of intelligence. It tells us of a mighty struggle that is now going on for a moral ascendency over the hearts of this world's population. It tell us that our race were seduced from their allegiance to God, by the plotting sagacity of one who stands pre-eminent against him, among the hosts of a very wide and extended rebellion. It tells us of the Captain of Salvation, who undertook to spoil him of this triumph; and throughout the whole of that magnificent train of prophecy which points to him, does it describe the work he had to do, as a conflict, in which strength was to be put forth, and painful suffering to be endured, and fury to be poured upon enemies, and principalities to be dethroned, and all those toils, and dangers, and difficulties to be borne, which strewed the path of perseverance that was to carry him to victory.

But it is a contest of skill, as well as of strength and of influence. There is the earnest competition of angelic faculties embarked on this struggle for ascendency. And while in the Bible there is recorded, (faintly and partially, we admit,) the deep and insidious policy that is practised on the one side; we are also told, that on the plan of our world's restoration, there are lavished all the riches of an unsearchable wisdom, upon the other. It would appear, that for the accomplishment of his purpose, the great enemy of God and of man plied his every calculation; and brought all the devices of his deep and settled malignity to bear upon our species; and thought that could he involve us in sin, every attribute of the Divinity stood staked to the banishment of our race from beyond the limits of the empire of righteousness; and thus did he practise his invasions on the moral territory of the unfallen; and glorying in his success, did he fancy and feel that he had achieved a permanent separation between the God who sitteth in heaven, and one at least of the planetary mansions which he had reared.

The errand of the Saviour was to restore this sinful world, and have its people re-admitted within the circle of heaven's pure and righteous family. But in the government of heaven, as well as in the government of earth, there are certain principles which cannot be compromised; and certain maxims of administration which must never be departed from; and a certain character of majesty and of truth, on which the taint even of the slightest violation can never be permitted; and a certain authority which must be upheld by the immutability of all its sanctions, and the unerring fulfilment of all its wise and righteous proclamations. All this was in the mind of the archangel, and a gleam of malignant joy shot athwart him, as he conceived his project for hemming our unfortunate species within the bound of an irrecoverable dilemma; and as surely as sin and holiness could not enter into fellowship, so surely did he think, that if man were seduced to disobedience, would the truth, and the justice, and the immutability of God, lay their insurmountable barriers on the path of his future acceptance.

It was only in that plan of recovery of which Jesus Christ was the author and the finisher, that the great adversary of our species met with a wisdom which over-matched him. It is true, that he had reared, in the guilt to which he seduced us, a mighty obstacle in the way of this lofty undertaking. But when the grand expedient was announced, and the blood of that atonement, by which sinners are brought nigh, was willingly offered to be shed for us, and the eternal Son, to carry this mystery into accomplishment, assumed our nature—then was the prince of that mighty rebellion, in which the fate and the history of our world are so deeply implicated, in visible alarm for the safety of all his acquisitions:—nor can the record of this wondrous history carry forward its narrative, without furnishing some transient glimpses of a sublime and a superior warfare, in which, for the prize of a spiritual dominion over our species, we may dimly perceive the contest of loftiest talent, and all the designs of heaven in behalf of man, met at every point of their evolution, by the counterworkings of a rival strength and a rival sagacity.

We there read of a struggle which the Captain of our salvation had to sustain, when the lustre of the Godhead lay obscured, and the strength of its omnipotence was mysteriously weighed down under the infirmities of our nature—how Satan singled him out, and dared him to the combat of the wilderness—how all his wiles and all his influences were resisted—how he left our Saviour in all the triumphs of unsubdued loyalty—how the progress of this mighty achievement is marked by the every character of a conflict—how many of the gospel miracles were so many direct infringements on the power and empire of a great spiritual rebellion—how in one precious season of gladness among the few which brightened the dark career of our Saviour's humiliation, he rejoiced in spirit, and gave as the cause of it to his disciples, that " he saw Satan fall like lightning from heaven"—how the momentary advantages that were gotten over him, are ascribed to the agency of this infernal being, who entered the heart of Judas, and tempted the disciple to betray his Master and his Friend. I know that I am treading on the confines of mystery. I cannot tell what the battle that he fought. I cannot compute the terror or the strength of his enemies. I cannot say, for I have not been told, how it was that they stood in marshalled and hideous array against him:—nor can I measure how great the firm daring of his soul, when he tasted that cup in all its bitterness, which he prayed might pass away from him; when with the feeling that he was forsaken by his God, he trod the wine-press alone; when he entered single-handed upon that dreary period of agony, and insult, and death, in which, from the garden to the cross, he had to bear the burden of a world's atonement. I cannot speak in my own language, but I can say, in the language of the Bible, of the days and the nights of this great enterprise, that it was the season of the travail of his soul; that it was the hour and the power of darkness; that the work of our redemption, was a work accompanied by the effort, and the violence, and the fury of a combat; by all the arduousness of a battle in its progress, and all the glories of a victory in its termination: and after he called out that it was finished, after he was loosed from the prison-house of the grave, after he had ascended up on high, he is said to have made captivity captive; and to have spoiled principalities and powers; and to have seen his pleasure upon his enemies; and to have made a show of them openly.

I will not affect a wisdom above that which is written, by fancying such details of this warfare as the Bible has not laid before me. But surely it is no more than being wise up to that which is written, to assert that in achieving the redemption of our world, a warfare had to be accomplished; that upon this subject there was among the higher provinces of creation, the keen and the animated conflict of opposing interests; that the result of it involved something grander and more affecting, than even the fate of this world's population; that it decided a question of rivalship between the righteous and everlasting Monarch of universal being, and the prince of a great and widely extended rebellion, of which I neither know how vast is the magnitude, nor how important and diversified are the bearings: and thus do we gather from this consideration, another distinct argument, helping us to explain, why on the salvation of our solitary species so much attention appears to have been concentred, and so much energy appears to have been expended.

But it would appear from the Records of Inspiration, that the contest is not yet ended; that on the one hand the Spirit of God is employed in making for the truths of Christianity, a way into the human heart, with all the power of an effectual demonstration; that on the other, there is a spirit now abroad, which worketh in the children of disobedience; that on the one hand, the Holy Ghost is calling men out of darkness into the marvellous light of the Gospel; and that on the other hand, he who is styled the god of this world, is blinding their hearts, lest the light of the glorious gospel of Christ should enter into them; that they who are under the dominion of the one, are said to have overcome, because greater is he that is in them than he that is in the world; and that they who are under the dominion of the other, are said to be the children of the devil, and to be under his snare, and to be taken captive by him at his will. How these respective powers do operate, is one question. The fact of their operation, is another. We abstain from the former. We attach ourselves to the latter, and gather from it, that the prince of darkness still walketh abroad amongst us; that he is still working his insidious policy, if not with the vigorous inspiration of hope, at least with the frantic energies of despair; that while the overtures of reconciliation are made to circulate through the world, he is plying all his devices to deafen and to extinguish the impression of them; or, in other words, while a process of invitation and of argument has emanated from heaven, for reclaiming men to their loyalty—the process is resisted at all its points, by one who is putting forth his every expedient, and wielding a mysterious ascendency, to seduce and to enthrall them.

To an infidel ear, all this carries the sound of something wild and visionary along with it. But though only known through the medium of revelation; after it is known, who can fail to recognise its harmony with the great lineaments of human experience? Who has not felt the workings of a rivalry within him, between the power of conscience and the power of temptation? Who does not remember those seasons of retirement, when the calculations of eternity had gotten a momentary command over the heart; and time, with all its interests and all its vexations, had dwindled into insignificancy before them? And who does not remember, how upon his actual engagement with the objects of time, they resumed a control, as great and as omnipotent, as if all the importance of eternity adhered to them—how they emitted from them such an impression upon his feelings, as to fix and to fascinate the whole man into a subserviency to their influence—how in spite of every lesson of their worthlessness, brought home to him at every turn by the rapidity of the seasons, and the vicissitudes of life, and the ever-moving progress of his own earthly career, and the visible ravages of death among his acquaintances around him, and the desolations of his family, and the constant breaking up of his system of friendships, and the affecting spectacle of all that lives and is in motion, withering and hastening to the grave;—oh! how comes it, that in the face of all this experience, the whole elevation of purpose, conceived in the hour of his better understanding, should be dissipated and forgotten? Whence the might, and whence the mystery of that spell, which so binds and so infatuates us to the world? What prompts us so to embark the whole strength of our eagerness and of our desires, in pursuit of interests which we know a few little years will bring to utter annihilation? Who is it that imparts to them all the charm and all the colour of an unfailing durability? Who is it that throws such an air of stability over these earthly tabernacles, as makes them look to the fascinated eye of man, like resting-places for eternity? Who is it that so pictures out the objects of sense, and so magnifies the range of their future enjoyment, and so dazzles the fond and deceived imagination, that in looking onward through our earthly career, it appears like the vista, or the perspective, of innumerable ages? He who is called the god of this world. He who can dress the idleness of its waking dreams in the garb of reality. He who can pour a seducing brilliancy over the panorama of its fleeting pleasures and its vain anticipations. He who can turn it into an instrument of deceitfulness; and make it wield such an absolute ascendency over all the affections, that man become the poor slave of its idolatries and its charms, puts the authority of conscience, and the warnings of the Word of God, and the offered instigations of the Spirit of God, and all the lessons of calculation, and all the wisdom even of his own sound and sober experience, away from him.

But, this wondrous contest will come to a close. Some will return to their loyalty, and others will keep by their rebellion; and, in the day of the winding up of the drama of this world's history, there will be made manifest to the myriads of the various orders of creation, both the mercy and vindicated majesty of the Eternal. Oh! on that day, how vain will this presumption of the Infidel astronomy appear, when the affairs of men come to be examined in the presence of an innumerable company; and beings of loftiest nature are seen to crowd around the judgment-seat; and the Saviour shall appear in our sky, with a celestial retinue, who have come with him from afar to witness all his doings, and to take a deep and solemn interest in all his dispensations; and the destiny of our species, whom the Infidel would thus detach, in solitary insignificance, from the universe altogether, shall be found to merge and to mingle with higher destinies—the good to spend their eternity with angels—the bad to spend their eternity with angels—the former to be re-admitted into the universal family of God's obedient worshippers—the latter to share in the everlasting pain and ignominy of the defeated hosts of the rebellious—the people of this planet to be implicated, throughout the whole train of their never-ending history, with the higher ranks, and the more extended tribes of intelligence: And thus it is, that the special administration we now live under, shall be seen to harmonise in its bearings, and to accord in its magnificence, with all that extent of nature and of her territories, which modern science has unfolded.

DISCOURSE VII.

ON THE SLENDER INFLUENCE OF MERE TASTE AND SENSIBILITY, IN MATTERS OF RELIGION.

"And, lo! thou art unto them as a very lovely song of one who hath a pleasant voice, and can play well on an instrument: for they hear thy words, but they do them not."

EZEKIEL xxxiii. 32.

YOU easily understand how a taste for music is one thing, and a real submission to the influence of religion is another—how the ear may be regaled by the melody of sound, and the heart may utterly refuse the proper impression of the sense that is conveyed by it—how the sons and daughters of the world may, with their every affection devoted to its perishable vanities, inhale all the delights of enthusiasm, as they sit in crowded assemblage around the deep and solemn oratorio—aye, and whether it be the humility of penitential feeling, or the rapture of grateful acknowledgement, or the sublime of a contemplative piety, or the aspiration of pure and of holy purposes, which breathes throughout the words of the performance, and gives to it all the spirit and all the expression by which it is pervaded; it is a very possible thing, that the moral, and the rational, and the active man, may have given no entrance into his bosom for any of these sentiments; and yet so overpowered may he be by the charm of the vocal conveyance through which they are addressed to him, that he may be made to feel with such an emotion, and to weep with such a tenderness, and to kindle with such a transport, and to glow with such an elevation, as may one and all carry upon them the semblance of sacredness.

But might not this semblance deceive him? Have you never heard any tell, and with complacency too, how powerfully his devotion was awakened by an act of attendance on the oratorio—how his heart, melted and subdued by the influence of harmony, did homage to all the religion of which it was the vehicle—how he was so moved and overborne, that he had to shed the tears of contrition, and to be agitated by the terrors of judgment, and to receive an awe upon his spirit of the greatness and the majesty of God—and that wrought up to the lofty pitch of eternity, he could look down upon the world, and by the glance of one commanding survey, pronounce upon the littleness and the vanity of all its concerns? Oh! it is very very possible that all this might thrill upon the ears of the man, and circulate a succession of solemn and affecting images around his fancy—and yet that essential principle of his nature, upon which the practical influence of Christianity turns, might have met with no reaching and no subduing efficacy whatever to arouse it. He leaves the exhibition, as dead in trespasses and sins as he came to it. Conscience has not wakened upon him. Repentance has not turned him. Faith has not made any positive lodgement within him of her great and her constraining realities. He speeds him back to his business and to his family, and there he plays off the old man in all the entireness of his uncrucified temper, and of his obstinate worldliness, and of all those earthly and unsanctified affections, which are found to cleave to him with as great tenacity as ever. He is really and experimentally the very same man as before—and all those sensibilities which seemed to bear upon them so much of the air and unction of heaven, are found to go into dissipation, and be forgotten with the loveliness of the song.

Amid all that illusion which such momentary visitations of seriousness and of sentiment throw around the character of man, let us never lose sight of the test, that "by their fruits ye shall know them." It is not coming up to this test, that you hear and are delighted. It is that you hear and do. This is the ground upon which the reality of your religion is discriminated now; and on the day of reckoning, this is the ground upon which your religion will be judged then; and that award is to be passed upon you, which will fix and perpetuate your destiny for ever. You have a taste for music. This no more implies the hold and the ascendency of religion over you, than that you have a taste for beautiful scenery, or a taste for painting, or even a taste for the sensualities of epicurism. But music may be made to express the glow and the movement of devotional feeling; and is it saying nothing, to say that the heart of him who listens with a raptured ear, is through the whole time of the performance, in harmony with such a movement? Why, it is saying nothing to the purpose. Music may lift the inspiring note of patriotism; and the inspiration may be felt; and it may thrill over the recesses of the soul, to the mustering up of all its energies; and it may sustain to the last cadence of the song, the firm nerve and purpose of intrepidity; and all this may be realised upon him, who in the day of battle, and upon actual collision with the dangers of it, turns out to be a coward. And music may lull the feelings into unison with piety; and stir up the inner man to lofty determinations; and so engage for a time his affections, that as if weaned from the dust, they promise an immediate entrance on some great and elevated career, which may carry him through his pilgrimage superior to all the sordid and grovelling enticements that abound in it. But he turns him to the world, and all this glow abandons him; and the words which he hath heard, he doeth them not; and in the hour of temptation he turns out to be a deserter from the law of allegiance; and the test I have now specified looks hard upon him, and discriminates him amid all the parading insignificance of his fine but fugitive emotions, to be the subject both of present guilt and of future vengeance.

The faithful application of this test would put to flight a host of other delusions. It may be carried round amongst all those phenomena of human character, where there is the exhibition of something associated with religion, but which is not religion itself. An exquisite relish for music is no test of the influence of Christianity. Neither are many other of the exquisite sensibilities of our nature. When a kind mother closes the eyes of her expiring babe, she is thrown into a flood of sensibility, and soothing to her heart are the sympathy and the prayers of an attending minister. When a gathering neighbourhood assemble to the funeral of an acquaintance, one pervading sense of regret and tenderness sits on the faces of the company;

and the deep silence, broken only by the solemn utterance of the man of God, carries a kind of pleasing religiousness along with it. The sacredness of the hallowed day, and all the decencies of its observation, may engage the affections of him who loves to walk in the footsteps of his father; and every recurring Sabbath may bring to his bosom, the charm of its regularity and its quietness. Religion has its accompaniments; and in these, there may be a something to soothe and to fascinate, even in the absence of the appropriate influences of religion. The deep and tender impression of a family-bereavement, is not religion. The love of established decencies, is not religion. The charm of all that sentimentalism which is associated with many of its solemn and affecting services, is not religion. They may form the distinct folds of its accustomed drapery; but they do not, any, or all of them put together, make up the substance of the thing itself. A mother's tenderness may flow most gracefully over the tomb of her departed little one; and she may talk the while of that heaven whither its spirit has ascended. The man whom death hath widowed of his friend, may abandon himself to the movements of that grief, which for a time will claim an ascendency over him; and, amongst the multitude of his other reveries, may love to hear of the eternity, where sorrow and separation are alike unknown. He who has been trained, from his infant days, to remember the Sabbath, may love the holiness of its aspect; and associate himself with all its observances; and take a delighted share in the mechanism of its forms. But, let not these think, because the tastes and the sensibilities which engross them, may be blended with religion, that they indicate either its strength or its existence within them. I recur to the test. I press its imperious exactions upon you. I call for fruit, and demand the permanency of a religious influence on the habits and the history. Oh! how many who take a flattering unction to their souls, when they think of their amiable feelings, and their becoming observations, with whom this severe touch-stone would, like the head of Medusa, put to flight all their complacency. The afflictive dispensation is forgotten—and he on whom it was laid, is practically as indifferent to God and to eternity as before. The Sabbath services come to a close; and they are followed by the same routine of week-day worldliness as before. In neither the one case nor the other, do we see more of the radical influence of Christianity, than in the sublime and melting influence of sacred music upon the soul; and all this tide of emotion is found to die away from the bosom, like the pathos or like the loveliness of a song.

The instances may be multiplied without number. A man may have a taste for eloquence, and eloquence the most touching or sublime may lift her pleading voice on the side of religion. A man may love to have his understanding stimulated by the ingenuities, or the resistless urgencies of an argument; and argument the most profound and the most overbearing, may put forth all the might of a constraining vehemence in behalf of religion. A man may feel the rejoicings of a conscious elevation, when some ideal scene of magnificence is laid before him; and where are these scenes so readily to be met with, as when led to expatiate in thought over the track of eternity, or to survey the wonders of creation, or to look to the magnitude of those great and universal interests which lie within the compass of religion. A man may have his attention riveted and regaled by that power of imitative description, which brings all the recollections of his own experience before him; which presents him with a faithful analysis of his own heart; which embodies in language such intimacies of observation and of feeling, as have often passed before his eyes, or played within his bosom, but had never been so truly or so ably pictured to the view of his remembrance. Now, all this may be done in the work of pressing the duties of religion; in the work of instancing the applications of religion; in the work of pointing those allusions to life and to manners, which manifest the truth to the conscience, and plant such a conviction of sin, as forms the very basis of a sinner's religion. Now, in all these cases, I see other principles brought into action, and which may be in a state of most lively and vigorous movement, and be yet in a state of entire separation from the principle of religion. I will make bold to say, on the strength of these illustrations, that as much delight may emanate from the pulpit, on an arrested audience beneath it, as ever emanated from the boards of a theatre—aye, and with as total a disjunction of mind too, in the one case as in the other, from the essence or the habit of religion. I recur to the test. I make my appeal to experience; and I put it to you all, whether your finding upon the subject do not agree with my saying about it, that a man may weep, and admire, and have many of his faculties put upon the stretch of their most intense gratification—his judgment established, and his fancy enlivened, and his feelings overpowered, and his hearing charmed as by the accents of heavenly persuasion, and all within him feasted by the rich and varied luxuries of an intellectual banquet!—Oh! it is cruel to frown unmannerly in the midst of so much satisfaction. But I must not forget that truth has her authority, as well as her sternness; and she forces me to affirm, that after all this has been felt and gone through, there might not be one principle which lies at the turning-point of conversion, that has experienced a single movement—not one of its purposes be conceived—not one of its doings be accomplished—not one step of that repentance, which, if we have not, we perish, so much as entered upon—not one announcement of that faith, by which we are saved, admitted into a real and actual possession by the inner man. He has had his hour's entertainment, and willingly does he award this homage to the performer, that he hath a pleasant voice, and can play well on an instrument—but, in another hour, it fleets away from his remembrance, and goes all to nothing, like the loveliness of a song.

Now, in bringing these Astronomical Discourses to a close, I feel it my duty to advert to this exhibition of character in man. The sublime and interesting topic which has engaged us, however feebly it may have been handled; however inadequately it may have been put in all its worth, and in all its magnitude before you; however short the representation of the speaker, or the conception of the hearers, may have been of that richness, and that greatness, and that loftiness, which belong to it; possesses in itself a charm to fix the attention, and to regale the imagination, and to subdue the whole man into a delighted reverence; and, in a word, to beget such a solemnity of thought and of emotion, as may occupy and enlarge the soul for hours together, as may waft it away from the grossness of ordinary life, and raise it to a kind of elevated calm above all its vulgarities and all its vexations.

Now, tell me whether the whole of this effect upon the feelings, may not be formed without the presence of religion. Tell me whether there might not be such a constitution of mind, that it may both want altogether that principle in virtue of which the doctrines of Christianity are admitted into the belief, and the duties of Christianity are admitted into a government over the practice—and yet at the very same time, it may have the faculty of looking abroad over some scene of magnificence, and of being wrought up to ecstacy with the sense of all those glories among which it is expatiating. I want you to see clearly the distinction between these two attributes of the human character. They are, in truth, as different the one from the other, as a taste for the grand and the graceful of scenery differs from the appetite of hunger; and the one may both exist and have a most intense operation within the bosom of that very individual, who entirely disowns, and is entirely disgusted with the other. What! must a man be converted, ere from the most elevated peak of some Alpine wilderness, he become capable of feeling the force and the majesty of those great lineaments which the hand of nature has thrown around him, in the varied forms of precipice, and mountain, and the wave of mighty forests, and the rush of sounding waterfalls, and distant glimpses of human territory, and pinnacles of everlasting snow, and the sweep of that circling horizon, which folds in its ample embrace the whole of this noble amphitheatre? Tell me whether without the aid of Christianity, or without a particle of reverence for the only name given under heaven whereby men can be saved, a man may not kindle at such a perspective as this, into all the raptures, and into all the movements of a poetic elevation; and be able to render into the language of poetry, the whole of that sublime and beauteous imagery which adorns it: aye, and as if he were treading on the confines of a sanctuary which he has not entered, may he not mix up with the power and the enchantment of his description, such allusions to the presiding genius of the scene; or to the still but animating spirit of the solitude; or to the speaking silence of some mysterious character which reigns throughout the landscape; or, in fine, to that eternal Spirit, who sits behind the elements he has formed, and combines them into all the varieties of a wide and a wondrous creation; might not all this be said and sung with an emphasis so moving, as to spread the colouring of piety over the pages of him who performs thus well upon his instrument; and yet, the performer himself have a conscience unmoved by a single warning of God's actual communication, and the judgment unconvinced, and the fears unawakened, and the life unreformed by it?

Now what is true of a scene on earth, is also true of that wider and more elevated scene which stretches over the immensity around it, into a dark and a distant unknown. Who does not feel an aggrandisement of thought and of faculty, when he looks abroad over the amplitudes of creation—when placed on a telescopic eminence, his aided eye can find a pathway to innumerable worlds—when that wondrous field, over which there had hung for many ages the mantle of so deep an obscurity, is laid open to him, and instead of a dreary and unpeopled solitude, he can see over the whole face of it such an extended garniture of rich and goodly habitations. Even the Atheist, who tells us that the universe is self-existent and indestructible—even he, who instead of seeing the traces of a manifold wisdom in its manifold varieties, sees nothing in them all but the exquisite structures and the lofty dimensions of materialism—even he, who would despoil creation of its God, cannot look upon its golden suns, and their accompanying systems, without the solemn impression of a magnificence that fixes and overpowers him. Now, conceive such a belief of God as you all profess, to dawn upon his understanding. Let him become as one of yourselves—and so be put into the condition of rising from the sublime of matter to the sublime of mind. Let him now learn to subordinate the whole of this mechanism to the design and authority of a great presiding Intelligence: and re-assembling all the members of the universe, however distant, into one family, let him mingle with his former conceptions of the grandeur which belonged to it, the conception of that eternal Spirit who sits enthroned on the immensity of his own wonders, and embraces all that he has made, within the ample scope of one great administration. Then will the images and the impressions of sublimity come in upon him from a new quarter. Then will another avenue be opened, through which a sense of grandeur may find its way into his soul, and have a mightier influence than ever to fill, and to elevate, and to expand it. Then will be established a new and a noble association, by the aid of which all that he formerly looked upon as fair, becomes more lovely; and all that he formerly looked upon as great, becomes more magnificent. But will you believe me, that even with this accession to his mind of ideas gathered from the contemplation of the Divinity; even with that pleasurable glow which steals over his imagination, when he now thinks him of the majesty of God; even with as much of what you would call piety, as I fear is enough to soothe and to satisfy many of yourselves, and which stirs and kindles within you when you hear the goings forth of the Supreme set before you in the terms of a lofty representation; even with all this, I say there may be as wide a distance from the habit and the character of godliness, as if God was still atheistically disowned by him. Take the conduct of his life and the currency of his affections; and you may see as little upon them of the stamp of loyalty to God, or of reverence for any one of his authenticated proclamations, as you may see in him who offers his poetic incense to the genii, or weeps enraptured over the visions of a beauteous mythology. The sublime of Deity has wrought up his soul to a pitch of conscious and pleasing elevation—and yet this no more argues the will of Deity to have a practical authority over him, than does that tone of elevation which is caught by looking at the sublime of a naked materialism. The one and the other have their little hour of ascendency over him; and when he turns him to the rude and ordinary world, both vanish alike from his sensibilities, as does the loveliness of a song.

To kindle and be elevated by a sense of the majesty of God, is one thing. It is totally another thing, to feel a movement of obedience to the will of God, under the impression of his rightful authority over all the creatures whom he has formed. A man may have an imagination all alive to the former; while the latter never prompts him to one act of obedience; never leads him to compare his life with the requirements of the Lawgiver; never carries him from such a scrutiny as this, to the conviction of sin; never whispers such an accusation to the ear of his conscience, as causes him to mourn, and to be in heaviness for the guilt of his hourly and habitual rebellion; never shuts him up to the conclusion of the need of a Saviour; never humbles him to acquiescence in the doctrine of that revelation, which comes to his door with such a host of evidence, as even his own philosophy cannot bid away; never extorts a single believing prayer in the name of Christ, or points a single look, either of trust or of reverence, to his atonement; never stirs any effective movement of conversion; never sends an aspiring energy into his bosom after the aids of that Spirit, who alone can waken him out of his lethargies, and by the anointing which remaineth, can rivet and substantiate in his practice, those goodly emotions which have hitherto plied him with the deceitfulness of their momentary visits, and then capriciously abandoned him.

The mere majesty of God's power and greatness, when offered to your notice, lays hold of one of the faculties within you. The holiness of God, with his righteous claim of legislation, lays hold of another of these faculties. The difference between them is so great, that the one may be engrossed and interested to the full, while the other remains untouched, and in a state of entire dormancy. Now, it is no matter what it be that ministers delight to the former of these two faculties: If the latter be not arrested and put on its proper exercise, you are making no approximation whatever to the right habit and character of religion. There are a thousand ways in which we may contrive to regale your taste for that which is beauteous and majestic. It may find its gratification in the loveliness of a vale, or in the freer and bolder outlines of an upland situation, or in the terrors of a storm, or in the sublime contemplations of astronomy, or in the magnificent idea of a God who sends forth the wakefulness of his omniscient eye, and the vigour of his upholding hand, throughout all the realms of nature and of providence. The mere taste of the human mind may get its ample enjoyment in each and in all of these objects, or in a vivid representation of them; nor does it make any material difference, whether this representation be addressed to you from the stanzas of a poem, or from the recitations of a theatre, or finally from the discourses and the demonstrations of a pulpit. And thus it is, that still on the impulse of the one principle only, people may come in gathering multitudes to the house of God; and share with eagerness in all the glow and bustle of a crowded attendance; and have their every eye directed to the speaker; and feel a responding movement in their bosom to his many appeals and his many arguments; and carry a solemn and overpowering impression of all the services away with them; and yet, throughout the whole of this seemly exhibition, not one effectual knock may have been given at the door of conscience. The other principle may be as profoundly asleep, as if hushed into the insensibility of death. There is a spirit of deep slumber, it would appear, which the music of no description, even though attuned to a theme so lofty as the greatness and majesty of the Godhead, can ever charm away. Oh! it may have been a piece of parading insignificance altogether—the minister playing on his favourite instrument, and the people dissipating away their time on the charm and idle luxury of a theatrical emotion.

The religion of taste, is one thing. The religion of conscience, is another. I recur to the test. What is the plain and practical doing which ought to issue from the whole of our argument? If one lesson come more clearly or more authoritatively out of it than another, it is the supremacy of the Bible. If fitted to impress one movement rather than another; it is that movement of docility, in virtue of which, man, with the feeling that he has all to learn, places himself in the attitude of a little child, before the book of the unsearchable God, who has deigned to break his silence, and to transmit even to our age of the world, a faithful record of his own communication. What progress then are you making in this movement? Are you, or are you not, like new born babes, desiring the sincere milk of the word, that you may grow thereby? How are you coming on in the work of casting down your lofty imaginations? With the modesty of true science, which is here at one with the humblest and most penitentiary feeling which Christianity can awaken, are you bending an eye of earnestness on the Bible, and appropriating its informations, and moulding your every conviction to its doctrines and its testimonies? How long, I beseech you, has this been your habitual exercise? By this time do you feel the darkness and the insufficiency of nature? Have you found your way to the need of an atonement? Have you learned the might and the efficacy which are given to the principle of faith? Have you longed with all your energies to realise it? Have you broken loose from the obvious misdoings of your former history? Are you convinced of your total deficiency from the spiritual obedience of the affections? Have you read of the Holy Ghost, by whom renewed in the whole desire and character of your mind, you are led to run with alacrity in the way of the commandments? Have you turned to its practical use, the important truth, that he is given to the believing prayers of all, who really want to be relieved from the power both of secret and of visible iniquity? I demand something more than the homage you have rendered to the pleasantness of the voice that has been sounded in your hearing. What I have now to urge upon you, is the bidding of the voice, to read, and to reform, and to pray, and, in a word, to make your consistent step from the elevations of philosophy, to all those exercises, whether of doing or of believing, which mark the conduct of the earnest, and the devoted, and the subdued, and the aspiring Christian.

This brings under our view, a most deeply

interesting exhibition of human nature, which may often be witnessed among the cultivated orders of society. When a teacher of Christianity addresses himself to that principle of justice within us, in virtue of which we feel the authority of God to be a prerogative which righteously belongs to him, he is then speaking the appropriate language of religion, and is advancing its naked and appropriate claim over the obedience of mankind. He is then urging that pertinent and powerful consideration, upon which alone he can ever hope to obtain the ascendency of a practical influence over the purposes and the conduct of human beings. It is only by insisting on the moral claim of God to a right of government over his creatures, that he can carry their loyal subordination to the will of God. Let him keep by this single argument, and urge it upon the conscience, and then, without any of the other accompaniments of what is called Christian oratory, he may bring convincingly home upon his hearers all the varieties of Christian doctrine. He may establish within their minds the dominion of all that is essential in the faith of the New Testament. He may, by carrying out this principle of God's authority into all its applications, convince them of sin. He may lead them to compare the loftiness and spirituality of his law, with the habitual obstinacy of their own worldly affections. He may awaken them to the need of a Saviour. He may urge them to a faithful and submissive perusal of God's own communication. He may thence press upon them the truth and the immutability of their Sovereign. He may work in their hearts an impression of this emphatic saying, that God is not to be mocked—that his law must be upheld in all the significancy of its proclamations—and that either its severities must be discharged upon the guilty, or in some other way an adequate provision be found for its outraged dignity, and its violated sanctions. Thus may he lead them to flee for refuge to the blood of the atonement. And he may farther urge upon his hearers, how, such is the enormity of sin, that it is not enough to have found an expiation for it; how its power and its existence must be eradicated from the hearts of all, who are to spend their eternity in the mansions of the celestial; how, for this purpose, an expedient is made known to us in the New Testament; how a process must be described upon earth, to which there is given the appropriate name of sanctification; how, at the very commencement of every true course of discipleship, this process is entered upon with a purpose in the mind of forsaking all; how nothing short of a single devotedness to the will of God, will ever carry us forward through the successive stages of this holy and elevated career; how, to help the infirmities of our nature, the Spirit is ever in readiness to be given to those who ask it; and that thus the life of every Christian becomes a life of entire dedication to him who died for us—a life of prayer, and vigilance, and close dependence on the grace of God—and, as the infallible result of the plain but powerful and peculiar teaching of the Bible, a life of vigorous unwearied activity in the doing of all the commandments.

Now, this I would call the essential business of Christianity. This is the truth as it is in Jesus, in its naked and unassociated simplicity. In the work of urging it, nothing more might have been done, than to present certain views, which may come with as great clearness, and freshness, and take as full possession of the mind of a peasant, as of the mind of a philosopher. There is a sense of God, and of the rightful allegiance that is due to him. There are plain and practical appeals to the conscience. There is a comparison of the state of the heart, with the requirements of a law which proposes to take the heart under its obedience. There is the inward discernment of its coldness about God; of its unconcern about the matters of duty and of eternity; of its devotion to the forbidden objects of sense; of its constant tendency to nourish within its own receptacles, the very element and principle of rebellion, and in virtue of this, to send forth the stream of an hourly and accumulating disobedience over those doings of the outer man, which make up his visible history in the world. There is such an earnest and overpowering impression of all this, as will fix a man down to the single object of deliverance; as will make him awake only to those realities which have a significant and substantial bearing on the case that engrosses him;

as will teach him to nauseate all the impertinences of tasteful and ambitious description; as will attach him to the truth in its simplicity; as will fasten his every regard upon the Bible, where, if he persevere in the work of honest inquiry, he will soon be made to perceive the accordancy between its statements, and all those movements of fear, or guilt, or deeply felt necessity, or conscious darkness, stupidity, and unconcern about the matters of salvation, which pass within his own bosom; in a word, as will endear to him that plainness of speech, by which his own experience is set evidently before him, and that plain phraseology of Scripture, which is best fitted to bring home to him the doctrine of redemption, in all the truth and in all the preciousness of its applications.

Now, the whole of this work may be going on, and that too in the wisest and most effectual manner, without so much as one particle of incense being offered to any of the subordinate principles of the human constitution. There may be no fascinations of style. There may be no magnificence of description. There may be no poignancy of acute and irresistible argument. There may be a riveted attention on the part of those whom the Spirit of God hath awakened to seriousness about the plain and affecting realities of conversion. Their conscience may be stricken, and their appetite be excited for an actual settlement of mind on those points about which they feel restless and unconfirmed. Such as these are vastly too much engrossed with the exigencies of their condition, to be repelled by the homeliness of unadorned truth. And thus it is, that while the loveliness of the song has done so little in helping on the influences of the gospel, our men of simplicity and prayer have done so much for it. With a deep and earnest impression of the truth themselves, they have made manifest that truth to the consciences of others. Missionaries have gone forth with no other preparation than the simple Word of the Testimony—and thousands have owned its power, by being both the hearers of the word and the doers of it also. They have given us the experiment in a state of unmingled simplicity; and we learn, from the success of their noble example, that without any one human expedient to charm the ear, the heart may, by the naked instrumentality of the Word of God, urged with plainness on those who feel its deceit and its worthlessness, be charmed to an entire acquiescence in the revealed way of God, and have impressed upon it the genuine stamp and character of godliness.

Could the sense of what is due to God, be effectually stirred up within the human bosom, it would lead to a practical carrying of all the lessons of Christianity. Now, to awaken this moral sense, there are certain simple relations between the creature and the Creator, which must be clearly apprehended, and manifested with power unto the conscience. We believe, that however much philosophers may talk about the comparative ease of forming those conceptions which are simple, they will, if in good earnest after a right footing with God, soon discover in their own minds, all that darkness and incapacity about spiritual things, which are so broadly announced to us in the New Testament. And oh! it is a deeply interesting spectacle, to behold a man, who can take a masterly and commanding survey over the field of some human speculation, who can clear his discriminated way through all the turns and ingenuities of some human argument, who by the march of a mighty and resistless demonstration, can scale with assured footstep the sublimities of science, and from his firm stand on the eminence he has won, can descry some wondrous range of natural or intellectual truth spread out in subordination before him:—and yet this very man, may, in reference to the moral and authoritative claims of the Godhead, be in a state of utter apathy and blindness! All his attempts, either at the spiritual discernment, or the practical impression of this doctrine, may be arrested and baffled by the weight of some great inexplicable impotency. A man of homely talents, and still homelier education, may see what he cannot see, and feel what he cannot feel; and wise and prudent as he is, there may lie the barrier of an obstinate and impenetrable concealment, between his accomplished mind, and those things which are revealed unto babes.

But while his mind is thus utterly devoid of

what may be called the main or elemental principle of theology, he may have a far quicker apprehension, and have his taste and his feelings much more powerfully interested, than the simple Christian who is beside him, by what may be called the circumstantials of theology. He can throw a wider and more rapid glance over the magnitudes of creation. He can be more delicately alive to the beauties and the sublimities which abound in it. He can, when the idea of a presiding God is suggested to him, have a more kindling sense of his natural majesty, and be able, both in imagination and in words, to surround the throne of the Divinity by the blazonry of more great, and splendid, and elevating images. And yet, with all those powers of conception which he does possess, he may not possess that on which practical Christianity hinges. The moral relation between him and God, may neither be effectively perceived, nor faithfully proceeded on. Conscience may be in a state of the most entire dormancy, and the man be regaling himself with the magnificence of God, while he neither loves God, nor believes God, nor obeys God.

And here I cannot but remark, how much effect and simplicity go together in the annals of Moravianism. The men of this truly interesting denomination, address themselves exclusively to that principle of our nature, on which the proper influence of Christianity turns. Or, in other words, they take up the subject of the gospel message, that message devised by him who knew what was in man, and who, therefore, knew how to make the right and the suitable application to man. They urge the plain Word of the Testimony; and they pray for a blessing from on high; and that thick impalpable veil, by which the god of this world blinds the hearts of men who believe not, lest the light of the glorious gospel of Christ should enter into them—that veil, which no power of philosophy can draw aside, gives way to the demonstration of the Spirit; and thus it is, that a clear perception of Scriptural truth, and all the freshness and permanency of its moral influences, are to be met with among men who have just emerged from the rudest and the grossest barbarity. Oh! when one looks at the number and the greatness of their achievements—when he thinks of the change they have made on materials so coarse and so unpromising—when he eyes the villages they have formed—and around the whole of that engaging perspective by which they have chequered and relieved the grim solitude of the desert, he witnesses the love, and listens to the piety of reclaimed savages;—who would not long to be in possession of the charm by which they have wrought this wondrous transformation—who would not willingly exchange for it all the parade of human eloquence, and all the confidence of human argument—and for the wisdom of winning souls, who is there that would not rejoice to throw the loveliness of the song, and all the insignificancy of its passing fascinations away from him?

And yet it is right that every cavil against Christianity should be met, and every argument for it be exhibited, and all the graces and sublimities of its doctrine be held out to their merited admiration. And if it be true, as it certainly is, that throughout the whole of this process, a man may be carried rejoicingly along from the mere indulgence of his taste, and the mere play and exercise of his understanding; while conscience is untouched, and the supremacy of moral claims upon the heart and the conduct is practically disowned by him—it is farther right that this should be adverted to; and that such a melancholy unhingement in the constitution of man should be fully laid open; and that he should be driven out of the seductive complacency which he is so apt to cherish, merely because he delights in the loveliness of the song; and that he should be urged with the imperiousness of a demand which still remains unsatisfied, to turn him from the corrupt indifference of nature, and to become personally a religious man; and that he should be assured how all the gratification he felt in listening to the word which respected the kingdom of God, will be of no avail, unless that kingdom come to himself in power—that it will only go to heighten the perversity of his character—that it will not extenuate his real and practical ungodliness, but will serve most fearfully to aggravate the condemnation of it.

With a religion so argumentable as ours, it

may be easy to gather out of it a feast for the human understanding. With a religion so magnificent as ours, it may be easy to gather out of it a feast for the human imagination. But with a religion so humbling, and so strict, and so spiritual, it is not easy to mortify the pride; or to quell the strong enmity of nature; or to arrest the currency of the affections; or to turn the constitutional habits; or to pour a new complexion over the moral history; or to stem the domineering influence of things seen and things sensible; or to invest faith with a practical supremacy; or to give its objects such a vivacity of influence as shall overpower the near and the hourly impressions, that are ever emanating upon man from a seducing world. It is here that man feels himself treading upon the limit of his helplessness. It is here that he sees where the strength of nature ends; and the power of grace must either be put forth, or leave him to grope his darkling way, without one inch of progress toward the life and the substance of Christianity. It is here that a barrier rises on the contemplation of the inquirer—the barrier of separation between the carnal and the spiritual, and on which he may idly waste the every energy which belongs to him, in the enterprise of surmounting it. It is here, that after having walked the round of nature's acquisitions, and lavished upon the truth all his ingenuities, and surveyed it in its every palpable character of grace and majesty, he will still feel himself on a level with the simplest and most untutored of the species. He needs the power of a living manifestation. He needs the anointing which remaineth. He needs that which fixes and perpetuates a stable revolution upon the character, and in virtue of which he may be advanced from the state of one who hears, and is delighted, to the state of one who hears, and is a doer. Oh! how strikingly is the experience even of vigorous and accomplished nature at one on this point with the announcements of revelation, that to work this change, there must be the putting forth of a peculiar agency; and that it is an agency, which, withheld from the exercise of loftiest talent, is often brought down on an impressed audience, through the humblest of all instrumentality, with the demonstration of the Spirit and with power.

Think it not enough, that you carry in your bosom an expanding sense of the magnificence of creation. But pray for a subduing sense of the authority of the Creator. Think it not enough, that with the justness of a philosophical discernment, you have traced that boundary which hems in all the possibilities of human attainment, and have found that all beyond it is a dark and fathomless unknown. But let this modesty of science be carried, as in consistency it ought, to the question of revelation, and let all the antipathies of nature be schooled to acquiescence in the authentic testimonies of the Bible. Think it not enough, that you have looked with sensibility and wonder at the representation of God throned in immensity, yet combining with the vastness of his entire superintendence, a most thorough inspection into all the minute and countless diversities of existence. Think of your own heart as one of these diversities; and that he ponders all its tendencies; and has an eye upon all its movements; and marks all its waywardness; and, God of judgment as he is, records its every secret, and its every sin, in the book of his remembrance. Think it not enough, that you have been led to associate a grandeur with the salvation of the New Testament, when made to understand that it draws upon it the regards of an arrested universe. How is it arresting your own mind? What has been the earnestness of your personal regards towards it? And tell me, if all its faith, and all its repentance, and all its holiness, are not disowned by you? Think it not enough, that you have felt a sentimental charm when angels were pictured to your fancy as beckoning you to their mansions, and anxiously looking to the every symptom of your grace and reformation. Oh! be constrained by the power of all this tenderness, and yield yourselves up in a practical obedience to the call of the Lord God merciful and gracious. Think it not enough, that you have shared for a moment in the deep and busy interest of that arduous conflict which is now going on for a moral ascendency over the species. Remember that the conflict is for each of you individually; and let this alarm you into a watchfulness against the power of every temptation, and a cleaving dependence upon him through whom alone you will be more than conquerors. Above all, forget not, that while you only hear and are delighted, you are still under nature's powerlessness and nature's condemnation—and that the foundation is not laid, the mighty and essential change is not accomplished, the transition from death unto life is not undergone, the saving faith is not formed, nor the passage taken from darkness to the marvellous light of the gospel, till you are both hearers of the word and doers also. "For if any be a hearer of the word and not a doer, he is like unto a man beholding his natural face in a glass: For he beholdeth himself, and goeth his way, and straightway forgetteth what manner of man he was."

APPENDIX.

The writer of these Discourses, has drawn up the following compilation of passages from Scripture, as serving to illustrate or to confirm the leading arguments which have been employed in each separate division of his subject.

DISCOURSE I.

In the beginning God created the heaven and the earth.—Gen. i. 1.

Thus the heavens and the earth were finished, and all the host of them.—Gen. ii. 1.

Behold the heaven, and the heaven of heavens, is the Lord's thy God, the earth also, with all that therein is.—Deut. x. 14.

There is none like unto the God of Jeshurun, who rideth upon the heaven in thy help, and in his excellency on the sky.—Deut. xxxiii. 26.

And Hezekiah prayed before the Lord, and said, O Lord God of Israel, which dwellest between the cherubims, thou art the God, even thou alone, of all the kingdoms of the earth; thou hast made heaven and earth.—2 Kings xix. 15.

For all the gods of the people are idols: but the Lord made the heavens.—1 Chronicles xvi. 26.

Thou, even thou, art Lord alone: thou hast made heaven, the heaven of heavens, with all their host, the earth and all things that are therein, the seas and all that is therein; and thou preservest them all; and the host of heaven worshippeth thee.—Nehemiah ix. 6.

Which alone spreadeth out the heavens, and treadeth upon the waves of the sea; which maketh Arcturus, Orion, and Pleiades, and the chambers of the south.—Job ix. 8, 9.

He stretcheth out the north over the empty place, and hangeth the earth upon nothing.—Job xxvi. 7.

By his spirit he hath garnished the heavens.—Job xxvi. 13.

The heavens declare the glory of God; and the firmament showeth his handy-work.—Psalm xix. 1.

By the word of the Lord were the heavens made; and all the host of them by the breath of his mouth.—Psalm xxxiii. 6.

Of old hast thou laid the foundation of the earth; and the heavens are the work of thy hands.—Psalm cii. 25.

Who coverest thyself with light as with a garment; who stretchest out the heavens like a curtain.—Psalm civ. 2.

He appointed the moon for seasons; the sun knoweth his going down.—Psalm civ. 19.

You are blessed of the Lord which made heaven and earth. The heaven, even the heavens, are the Lord's: but the earth hath he given to the children of men.—Psalm cxv. 15, 16.

My help cometh from the Lord, which made heaven and earth.—Psalm cxxi. 2.

Our help is in the name of the Lord, who made heaven and earth.—Psalm cxxiv. 8.

The Lord that made heaven and earth, bless thee out of Zion.—Psalm cxxxiv. 3.

Which made heaven and earth, the sea, and all that therein is.—Psalm cxlvi. 6.

The Lord by wisdom hath founded the earth; by understanding hath he established the heavens.—Prov. iii. 19.

Who hath measured the waters in the hollow of his hand, and meted out heaven with the span, and comprehended the dust of the earth in a measure, and weighed the mountains in a scale, and the hills in a balance.—Isa. xl. 12.

It is he that sitteth upon the circle of the earth, and the inhabitants thereof are as grasshoppers; that stretcheth out the heavens as a curtain, and spreadeth them out as a tent to dwell in.—Isa. xl. 22.

Thus saith God the Lord, he that created the heavens, and stretched them out; he that spread forth the earth, and that which cometh out of it; he that giveth breath unto the people upon it, and spirit to them that walk therein.—Isa. xlii. 5.

Thus saith the Lord, thy Redeemer, and he that formed thee from the womb, I am the Lord that maketh all things; that stretcheth forth the heavens alone; that spreadeth abroad the earth by myself.—Isa. xliv. 24.

I have made the earth, and created man upon it; I, even my hands, have stretched out the heavens, and all their host have I commanded.—Isa. xlv. 12.

For thus saith the Lord that created the heavens, God himself that formed the earth and made it, he hath established it, he created it not in vain, he formed it to be inhabited.—Isa. xlv. 18.

Mine hand also hath laid the foundation of the earth, and my right hand hath spanned the heavens; when I call unto them, they stand up together.—Isa. xlviii. 13.

He hath made the earth by his power, he hath established the world by his wisdom, and hath stretched out the heavens by his discretion.—Jer. x. 12.

Ah Lord God! behold, thou hast made the heaven and the earth by thy great power and stretched out arm, and there is nothing too hard for thee.—Jer. xxxii. 17.

He hath made the earth by his power, he hath established the world by his wisdom, and hath stretched out the heaven by his understanding.—Jer. li. 15.

It is he that buildeth his stories in the heaven, and hath founded his troop in the earth; he that calleth for the waters of the sea, and poureth them out upon the face of the earth, The Lord is his name.—Amos ix. 6.

We also are men of like passions with you, and preach unto you, that ye should turn from these vanities unto the living God, which made heaven, and earth, and the sea, and all things that are therein.—Acts xiv. 15.

Hath in these last days spoken unto us by his Son, whom he hath appointed heir of all things, by whom also he made the worlds.—Heb. i. 2.

Thou, Lord, in the beginning hast laid the foundation of the earth; and the heavens are the work of thine hands.—Heb. i. 10.

Through faith, we understand that the worlds were framed by the word of God.—Heb. xi. 3.

DISCOURSE II.

The secret things belong unto the Lord our God, but those things which are revealed belong unto us and to our children for ever, that we may do all the words of this law.—Deut. xxix. 29.

I would seek unto God, and unto God would I commit my cause; Which doeth great things and unsearchable; marvellous things without number.—Job v. 8, 9.

Which doeth great things past finding out; yea, and wonders without number.—Job ix. 10.

Canst thou by searching find out God? canst thou find out the Almighty unto perfection?—Job xi. 7.

Hast thou heard the secret of God? and dost thou restrain wisdom to thyself?—Job xv. 8.

Lo, these are parts of his ways; but how little a portion is heard of him? but the thunder of his power who can understand?—Job xxvi. 14.

Behold, God is great, and we know him not; neither can the number of his years be searched out.—Job. xxxvi. 26.

God thundereth marvellously with his voice; great things doeth he, which we cannot comprehend.—Job xxxvii. 5.

Touching the Almighty, we cannot find him out: he is excellent in power, and in judgment, and in plenty of justice.—Job xxxvii. 23.

Thy way is in the sea, and thy path in the great waters, and thy footsteps are not known.—Psalm lxxvii. 19.

Great is the Lord, and greatly to be praised; and his greatness is unsearchable.—Psalm cxlv. 3.

For my thoughts are not your thoughts, neither are your ways my ways, saith the Lord. For as the heavens are higher than the earth, so are my ways higher than your ways, and my thoughts than your thoughts.—Isa. lv. 8, 9.

Verily I say unto you, except ye be converted, and become as little children, ye shall not enter into the kingdom of heaven.—Matth. xviii. 3.

Verily I say unto you, whosoever shall not receive the kingdom of God as a little child, shall in no wise enter therein.—Luke xviii. 17.

O the depth of the riches both of the wisdom and knowledge of God! how unsearchable are his judgments, and his ways past finding out! For who hath known the mind of the Lord? Or who hath been his counsellor?—Rom. xi. 33, 34.

Let no man deceive himself. If any man among you seemeth to be wise in this world, let him become a fool, that he may be wise.—1 Cor. iii. 18.

For if a man thinketh himself to be something, when he is nothing, he deceiveth himself.—Gal. vi. 3.

Beware lest any man spoil you through philosophy and vain deceit, after the tradition of men, after the rudiments of the world, and not after Christ.—Col. ii. 8.

O Timothy, keep that which is committed to thy trust, avoiding profane and vain babblings, and oppositions of science falsely so called.—1 Tim. vi. 20.

DISCOURSE III.

But will God indeed dwell on the earth? Behold, the heaven, and heaven of heavens, cannot contain thee; how much less this house that I have builded? Yet have thou respect unto the prayer of thy servant, and to his supplication, O Lord my God, to hearken unto the cry and to the prayer which thy servant prayeth before thee to-day: That thine eyes may be open toward this house night and day, even toward the place of which thou hast said, My name shall be there; that thou mayest hearken unto the prayer which thy servant shall make toward this place.—1 Kings viii. 27, 28, 29.

For he looketh to the ends of the earth, and seeth under the whole heaven.—Job xxviii. 24.

For his eyes are upon the ways of man, and he seeth all his goings.—Job xxxiv. 21.

Though the Lord be high, yet hath he respect unto the lowly.—Psalm cxxxviii. 6.

O Lord thou hast searched me and known me. Thou knowest my down-sitting and mine up-rising: thou understandest my thought afar off. Thou compassest my path and my lying down, and art acquainted with all my ways. For there is not a word in my tongue, but lo, O Lord! thou knowest it altogether. Thou hast beset me behind and before, and laid thine hand upon me. Such knowledge is too wonderful for me; it is high, I cannot attain unto it. Whither shall I go from thy Spirit, or whither shall I flee from thy presence?—Psalm cxxxix. 1—7.

How precious also are thy thoughts unto me, O God! how great is the sum of them! If I should count them, they are more in number than the sand: when I awake, I am still with thee.—Psalm cxxxix. 17, 18.

The eyes of the Lord are in every place, beholding the evil and the good.—Prov. xv. 3.

Can any hide himself in secret places, that I shall not see him? saith the Lord: do not I fill heaven and earth? saith the Lord.—Jer. xxiii. 24.

Behold the fowls of the air: for they sow not, neither do they reap, nor gather into barns; yet your heavenly Father feedeth them. Are ye not much better than they? And why take ye thought for raiment? Consider the lilies of the field how they grow; they toil not, neither do they spin: And yet I say unto you, That even Solomon, in all his glory, was not arrayed like one of these. Wherefore, if God so clothe the grass of the field, which to-day is, and to-morrow is cast into the oven, shall he not much more clothe you, O ye of little faith?—Matth. vi. 26. 28, 29, 30.

But the very hairs of your head are all numbered.—Matth. x. 30.

Neither is there any creature that is not manifest in his sight: but all things are naked and opened unto the eyes of him with whom we have to do.—Heb. iv. 13.

DISCOURSE IV.

And he dreamed, and behold a ladder set up on the earth, and the top of it reached to heaven: and behold the angels of God ascending and descending on it.—Gen. xxviii. 12.

For a thousand years in thy sight, are but as yesterday when it is past, and as a watch in the night.—Psalm xc. 4.

Lift up your eyes to the heavens, and look upon the earth beneath: for the heavens shall vanish away like smoke, and the earth shall wax old like a garment, and they that dwell therein, shall die in like manner; but my salvation shall be for ever, and my righteousness shall not be abolished.—Isa. li. 6.

For the Son of Man shall come in the glory of his Father with his angels; and then he shall reward every man according to his works.—Matth. xvi. 27.

When the Son of Man shall come in his glory, and all the holy angels with him, then shall he sit upon the throne of his glory.—Matth. xxv. 31.

Also I say unto you, Whosoever shall confess me before men, him shall the Son of Man also confess before the angels of God. But he that denieth me before men, shall be denied before the angels of God.—Luke xii. 8, 9.

And he saith unto him, Verily, verily, I say unto you, hereafter ye shall see heaven open, and the angels of God ascending and descending upon the Son of Man.—John i. 51.

We are made a spectacle to the world, and to angels, and to men.—1 Cor. iv. 9.

Wherefore God also hath highly exalted him, and given him a name which is above every name. That at the name of Jesus every knee should bow, of things in heaven and things in earth, and things under the earth; And that every tongue should confess that Jesus Christ is Lord, to the glory of God the Father.—Phil. ii. 9, 10, 11.

When the Lord Jesus shall be revealed from heaven with his mighty angels.—2 Thess. i. 7.

And without controversy great is the mystery of godliness: God was manifest in the flesh, justified in the Spirit, seen of angels, preached unto the Gentiles, believed on in the world, received up into glory.—1 Tim. iii. 16.

I charge thee before God, and the Lord Jesus Christ, and the elect angels, that thou observe these things.—1 Tim. v. 21.

And again when he bringeth in the first-begotten into the world, he saith, And let all the angels of God worship him. —Heb. i. 6.

But ye are come unto Mount Zion, and unto the city of the living God, the heavenly Jerusalem, and to an innumerable company of angels, To the general assembly and church of the first-born, which are written in heaven, and to God the judge of all, and to the spirits of just men made perfect, And to Jesus, the Mediator of the new covenant.—Heb. xii. 22, 23, 24.

But, beloved be not ignorant of this one thing, that one day is with the Lord as a thousand years, and a thousand years as one day. The Lord is not slack concerning his promise, as some men count slackness; but is long-suffering to us-ward, not willing that any should perish, but that all should come to repentance. But the day of the Lord will come as a thief in the night; in the which the heavens shall pass away with a great noise, and the elements shall melt with fervent heat, the earth also and the works that are therein, shall be burnt up.—2 Peter iii. 8, 9, 10.

And the angel which I saw stand upon the sea and upon the earth, lifted up his hand to heaven, And sware by him that liveth for ever and ever, who created heaven and the things that therein are, and the earth and the things that therein are, and the sea and the things which are therein, that there should be time no longer.—Rev. x. 5, 6.

And the third angel followed them, saying with a loud voice, If any man worship the beast and his image, and receive his mark in his forehead or in his hand, The same shall drink of the wine of the wrath of God, which is poured out without mixture into the cup of his indignation; and he shall be tormented with fire and brimstone in the presence of the holy angels, and in the presence of the Lamb.—Rev. xiv. 9, 10.

And I saw a great white throne, and him that sat on it, from whose face the earth and the heaven fled away, and there was found no place for them.—Rev. xx. 11.

DISCOURSE V.

And Nathan departed unto his house: and the Lord struck the child that Uriah's wife bare unto David, and it was very sick. David therefore besought God for the child: and David fasted, and went in and lay all night upon the earth.

And the elders of his house arose, and went to him, to raise him up from the earth: but he would not, neither did he eat bread with them. And it came to pass on the seventh day, that the child died. And the servants of David feared to tell him that the child was dead; for they said, Behold, while the child was yet alive, we spake unto him, and he would not hearken unto our voice; how will he then vex himself, if we tell him that the child is dead? But when David saw that his servants whispered, David perceived that the child was dead: therefore David said unto his servants, Is the child dead? And they said, he is dead. Then David arose from the earth, and washed, and anointed himself, and changed his apparel, and came into the house of the Lord, and worshipped: then he came to his own house: and, when he required, they set bread before him, and he did eat. Then said his servants unto him. What thing is this that thou hast done? Thou didst fast and weep for the child while it was alive: but when the child was dead, thou didst rise and eat bread. And he said, while the child was yet alive, I fasted and wept: for I said, who can tell whether God will be gracious to me, that the child may live? But now he is dead, wherefore should I fast? can I bring him back again? I shall go to him, but he shall not return to me.—2 Sam. xii. 15—23.

The angel of the Lord encampeth round about them that fear him, and delivereth them.—Psalm xxxiv. 7.

For he shall give his angels charge over thee, to keep thee in all thy ways.—Psalm xci. 2.

And he shall send his angels with a great sound of a trumpet; and they shall gather together his elect from the four winds, from one end of heaven to the other.—Matth. xxiv. 31.

Likewise, I say unto you, There is joy in the presence of the angels of God over one sinner that repenteth.—Luke xv. 10.

Are they not all ministering spirits, sent forth to minister for them who shall be heirs of salvation?—Heb. i. 14.

DISCOURSE VI.

Then was Jesus led up of the Spirit into the wilderness, to be tempted of the devil.—Matth. iv. 1.

The enemy that sowed them is the devil; the harvest is the end of the world; and the reapers are the angels. The Son of Man shall send forth his angels, and they shall gather out of his kingdom all things that offend, and them which do iniquity.—Matth. xiii. 39. 41.

Then shall he say also unto them on the left hand, Depart from me, ye cursed, into everlasting fire, prepared for the devil and his angels.—Matth. xxv. 41.

And in the synagogue there was a man which had a spirit of an unclean devil, and cried out with a loud voice, saying, Let us alone; what have we to do with thee, thou Jesus of Nazareth? art thou come to destroy us? I know thee who thou art: the Holy One of God.—Luke iv. 33, 34.

Those by the way-side are they that hear; then cometh the devil, and taketh away the word out of their hearts, lest they should believe and be saved.—Luke viii. 12.

But he, knowing their thoughts, said unto them, Every kingdom divided against itself is brought to desolation; and a house divided against a house, falleth. If Satan also be divided against himself, how shall his kingdom stand? because ye say that I cast out devils through Beelzebub.—Luke xi. 17, 18.

Ye are of your father the devil, and the lusts of your father ye will do: he was a murderer from the beginning, and abode not in the truth, because there is no truth in him. When he speaketh a lie, he speaketh of his own: for he is a liar, and the father of it.—John viii. 44.

And supper being ended, (the devil having now put into the heart of Judas Iscariot, Simon's son, to betray him.)—John xiii. 2.

But Peter said, Ananias, why hath Satan filled thine heart to lie to the Holy Ghost, and to keep back part of the price of the land.—Acts v. 3.

To open their eyes, and to turn them from darkness to light, and from the power of Satan unto God, that they may receive forgiveness of sins, and inheritance among them which are sanctified by faith that is in me.—Acts xxvi. 18.

And the God of peace shall bruise Satan under your feet shortly. The grace of our Lord Jesus Christ be with you. Amen.—Rom. xvi. 20.

Lest Satan should get an advantage of us: for we are not ignorant of his devices.—2 Cor. ii. 11.

In whom the god of this world hath blinded the minds of them which believe not, lest the light of the glorious gospel of Christ, who is the image of God, should shine unto them. 2 Cor. iv. 4.

Wherein in time past ye walked according to the course of this world, according to the prince of the power of the air, the spirit that now worketh in the children of disobedience.—Eph. ii. 2.

Put on the whole armour of God, that ye may be able to stand against the wiles of the devil. For we wrestle not against flesh and blood, but against principalities, against powers, against the rulers of the darkness of this world, against spiritual wickedness in high places.—Eph. vi. 11, 12.

For some are already turned aside after Satan.—1 Tim. v. 15.

Forasmuch then as the children are partakers of flesh and blood, he also himself likewise took part of the same; that through death he might destroy him that had the power of death, that is the devil.—Heb. ii. 14.

Submit yourselves therefore to God. Resist the devil, and he will flee from you.—James iv. 7.

Be sober, be vigilant; because your adversary the devil, as a roaring lion, walketh about, seeking whom he may devour: Whom resist, steadfast in the faith, knowing that the same afflictions are accomplished in your brethren that are in the world.—1 Pet. v. 8, 9.

He that committeth sin, is of the devil; for the devil sinneth from the beginning. For this purpose the Son of God was manifested, that he might destroy the works of the devil.

In this the children of God are manifest, and the children of the devil: whosoever doeth not righteousness is not of God, neither he that loveth not his brother.—1 John iii. 8. 10.

Ye are of God, little children, and have overcome them; because greater is he that is in you, than he that is in the world.—1 John iv. 4.

And the angels which kept not their first estate, but left their own habitation, he hath reserved in everlasting chains, under darkness, unto the judgment of the great day.—Jude 6.

He that overcometh, the same shall be clothed in white raiment; and I will not blot out his name out of the book of life, but I will confess his name before my Father, and before his angels.—Rev. iii. 5.

And there was war in heaven: Michael and his angels fought against the dragon; and the dragon fought and his angels, And prevailed not; neither was their place found any more in heaven. And the great dragon was cast out. that old serpent, called the Devil, and Satan, which deceiveth the whole world; he was cast out into the earth, and his angels were cast out with him. Therefore rejoice, ye heavens, and ye that dwell in them. Woe to the inhabiters of the earth and of the sea! for the devil is come down unto you, having great wrath, because he knoweth that he hath but a short time.—Rev. xii. 7, 8, 9. 12.

And he laid hold on the dragon, that old serpent, which is the Devil, and Satan, and bound him a thousand years, And when the thousand years are expired, Satan shall be loosed out of his prison. And the devil that deceived them was cast into the lake of fire and brimstone, where the beast and the false prophet are, and shall be tormented day and night, for ever and ever.—Rev. xx. 2, 7, 10.

DISCOURSE VII.

Therefore, whosoever heareth these sayings of mine, and doeth them, I will liken him unto a wise man, which built his house upon a rock: And the rain descended, and the floods came, and the winds blew, and beat upon that house; and it fell not: for it was founded upon a rock. And every one that heareth these sayings of mine, and doeth them not, shall be likened unto a foolish man, which built his house upon the sand: And the rain descended, and the floods came, and the winds blew, and beat upon that house; and it fell: and great was the fall of it.—Matth. vii. 24—27.

At that time, Jesus answered and said, I thank thee, O Father! Lord of heaven and earth, because thou hast hid these things from the wise and prudent, and hast revealed them unto babes.—Matth. xi. 25.

Then shall ye begin to say, We have eaten and drank in thy presence, and thou hast taught in our streets. But he shall say, I tell you, I know you not whence ye are: depart from me all ye workers of iniquity.—Luke xiii. 26, 27.

For not the hearers of the law are just before God, but the doers of the law shall be justified.—Rom. ii. 13.

And I, brethren, when I came to you, came not with excellency of speech or of wisdom, declaring unto you the testimony of God. For I determined not to know any thing among you, save Jesus Christ and him crucified. And my speech and my preaching was not with enticing words of man's wisdom, but in demonstration of the Spirit and of power. That your faith should not stand in the wisdom of men, but in the power of God. Now we have received not the spirit of the world, but the Spirit which is of God; that we might know the things that are freely given to us of God. Which things also we speak, not in the words which man's wisdom teacheth, but which the Holy Ghost teacheth; comparing spiritual things with spiritual. But the natural man receiveth not the things of the Spirit of God; for they are foolishness unto him: neither can he know them, because they are spiritually discerned.—1 Cor. ii. 1, 2. 4, 5. 12, 13, 14.

For the wisdom of this world is foolishness with God.—1 Cor. iii. 19.

For the kingdom of God is not in word, but in power.—1 Cor. iv. 20.

Forasmuch as ye are manifestly declared to be the epistle of Christ ministered by us, written not with ink, but with the Spirit of the living God; not in tables of stone, but in fleshly tables of the heart. Not that we are sufficient of ourselves to think any thing as of ourselves; but our sufficiency is of God; Who also hath made us able ministers of the New Testament; not of the letter, but of the spirit: for the letter killeth, but the spirit giveth life.—2 Cor. iii. 3. 5, 6.

That the God of our Lord Jesus Christ, the Father of glory, may give unto you the spirit of wisdom and revelation in the knowledge of him: The eyes of your understanding being enlightened; that ye may know what is the hope of his calling, and what the riches of the glory of his inheritance in the saints, And what is the exceeding greatness of his power to us-ward who believe, according to the working of his mighty power.—Eph. i. 17, 18, 19.

And you hath he quickened, who were dead in trespasses and sins. For we are his workmanship, created in Christ Jesus unto good works.—Eph. ii. 1. 10.

For our gospel came not unto you in word only, but also in power, and in the Holy Ghost, and in much assurance.—1 Thes. i. 5.

Of his own will begat he us with the word of truth, that we should be a kind of first-fruits of his creatures.

But be ye doers of the word, and not hearers only, deceiving yourselves. For if any be a hearer of the word, and not a doer, he is like unto a man beholding his natural face in a glass. For he beholdeth himself, and goeth his way, and straightway forgetteth what manner of man he was. But whoso looketh into the perfect law of liberty, and continueth therein, he being not a forgetful hearer, but a doer of the work, this man shall be blessed in his deed.—James i. 18. 22—25.

But ye are a chosen generation, a royal priesthood, an holy nation, a peculiar people, that ye should show forth the praises of him who has called you out of darkness into his marvellous light.—1 Pet. ii. 9.

But ye have an unction from the Holy One, and ye know all things.

But the anointing which ye have received of him abideth in you: and ye need not that any man teach you: but as the same anointing teacheth you of all things, and is truth, and is no lie, and even as it hath taught you, ye shall abide in him.—1 John ii. 20. 27.

THE END.

Printed in the United States
By Bookmasters